OUR COSMIC STORY

Exploring Life, Civilization, and the Universe

by

MATHEW ANDERSON

Mathew C. Anderson

OUR COSMIC STORY
Exploring Life, Civilization, and the Universe

Cover art by Mathew Anderson
Robot art by Freepik

Published by Amazon
www.amazon.com/author/mathewanderson

Additional copies available at:
www.ourcosmicstory.com

Our Cosmic Story on Twitter and Facebook @OurCosmicStory

Printed and bound in the United States of America
by Amazon: www.amazon.com/author/mathewanderson

For All Great Explorers

"Equipped with his five senses, man explores the Universe around him and calls the adventure Science." – Edwin Hubble

ACKNOWLEDGMENTS

The idea and motivation to start writing this book came from a personal quest to understand the Universe and our special place on this pale blue dot called Earth. I felt the attempt must be made to gain a better understanding of how it all came together, what the chances are for it to do the same elsewhere, but perhaps just in a little bit different of a way to be interesting to discover.

I always say that writing a book is just a third of the battle, with editing making up the bulk of the work. There have been more drafts of the book than I can count. The more valued contributions have come from family, friends, and other contributors. I am thankful for the patience of everyone, especially those who let me pester them to read a chapter or look over a phrase, often when the subject was not yet fully formed on paper, but just a drifting image in my mind. Writing a book is often seen as a solitary affair, but it has been anything but that in my experience.

For all editing efforts, I am thankful to many individuals, including but not limited to: Ammy Sriyunyongwat, Ben Roye, Chuck Bird, Cindy Anderson, Cora Nelson, David Zhong, Jason Searcy, Jessica Anderson, Jennifer Norian, Josh Maida, Michael Rogers, Richard Garriott, Scott Jennings, Starr Long, Tin Khuong, Tony Medrano, as well as the thousands of my social-network friends and acquaintances!

CONTENTS

OUR COSMIC STORY

Exploring Life, Civilization, and the Universe

INTRODUCTION – A BIG PICTURE VIEW

"When scientists are asked what they are working on, their response is seldom 'Finding the origin of the Universe' or 'Seeking to cure cancer.' Usually, they will claim to be tackling a very specific problem - a small piece of the jigsaw that builds up the big picture." - Martin Rees

Precious little time exists where we get a chance to reflect upon life, to smell the roses or to look at the cosmic picture, yet it is important to find the time for such reflection when the grand Universe is on our doorstep. How did life get started? How did our fantastically complex civilization develop? When will we encounter other civilizations out there, if ever? This book will explore our history and place in the Universe, examine why Earth is so hospitable for life and civilization, and consider the likelihood for life to exist on other worlds, some that may be far more different than our own.

The quest to reach beyond the confines of our world is a natural consequence of being a very small part of a grand and dynamic Universe. Looking up at the sky instills within us some expectation that we are not alone, and we wonder if there is not something amazing happening out there somewhere. This sense of awe may not be exclusive to Earthlings; for in a galaxy truly far away, there could be creatures with similar musings as they peer towards our corner of the Universe. The idea that we share a common experience with others in the cosmos is something to keep in mind while reading this book.

Regardless of the possibility of countless other life forms existing on rocky barges adrift in the cosmos, we should still hold the belief that humanity is special. Our world is rare enough that we may appreciate it just as much as if we are in fact alone. Recent studies have shown that, while life's ingredients are common throughout the Universe, the exact quantity and assortment of chemicals and minerals that make up Earth are unlikely to exist elsewhere. This may have significant consequences for life and evolution on another world that's close in properties, but not quite the same as Earth.

For a close-to-home example, both Mars and Venus can be considered distant cousins of Earth and once thought to be habitable, though you wouldn't want to book a vacation to either of them anytime soon. While Mars and Venus are rocky bodies with solid cores, Mars today lacks a dense oxygen-rich atmosphere and liquid water on its surface because the planet is simply too small. Venus is nearly the same size as Earth, but is far too close to our scorching Sun. There are many other variables of habitability to consider as well, many of which we will explore in upcoming chapters. The physical makeup of Earth and how it compares to other planets in the solar system is an important starting point for understanding where we may find life elsewhere in the Universe.

Even though we may currently be well off on Earth, humanity should make every effort to voyage into space, if only for the practical reason that our planet will not support life forever. We may one day need to flee Earth in order to preserve the existence of our species from a variety of cosmic and terrestrial extinction events, like asteroid impacts and supervolcanoes. Currently, all of our proverbial eggs are dangerously in one basket. While the lack of evidence of life on other worlds may suggest at first that the endeavor to colonize space is futile, this is a dangerous assumption we just cannot afford to make.

The act of colonizing space will of course come at initially great cost, but in the long run it may pay off in ways we cannot even imagine. Many great explorers like Marco Polo, Sir Francis Drake, Ferdinand Magellan, and Charles Darwin risked their lives and boldly faced peril to sail across vast oceans in the name of science and discovery. Diseases and other formidable barriers attempted to stop them from making progress, but they pushed on in the hope that a better future lay ahead. These great explorers, and many great thinkers throughout history, have helped to lay the foundation our civilization stands on today.

Whatever our future may hold, let us not forget where we came from or fail to cherish our home. Protecting Earth for as long as possible may be important for facilitating our ability to leave it someday – and perhaps in our need to return to it, should space be more unforgiving than we had realized. Our planet is not a place that we can spoil with the assumption that a better one will eventually be found. Assuming that our long-term existence is assured will be the ultimate undoing of our way of life. Many of Earth's past civilizations made this arrogant assumption about their destinies, and it resulted in their swift demise.

In order to begin to understand how humanity has achieved all that we have so far, and further our chances of carrying our knowledge into space, there is one thing that must always be with us: a collective sense of hope and drive to improve the whole of society. The motivation to pass down prosperity to future generations has the power to be a catalyst for expanding into a spacefaring civilization that can counter the constant threats against our world. While great things will still be accomplished if we stay grounded to Earth, it would be tragic if future generations one day forget that we once long ago nearly made it to the stars, but did not.

I hope that this book heightens your sense of wonder about our tiny but special place in the cosmos, as well as fires your imagination and intensifies your intrigue in exploring the potential for other worlds where we may one day call home.

COMMON DEFINITIONS

The following definitions can be found to vary, so I chose ones that are fit for the book's subject matter. Definitions are either from or based on the Oxford English Dictionary (OED), NASA.gov, Space.com, or other official sources:

Civilization: The process by which a society or place reaches an advanced stage of social development and organization.

Convection: The movement caused within a fluid by the tendency of hotter and therefore less dense material to rise, and colder, denser material to sink under the influence of gravity, which consequently results in transfer of heat.

Cosmos/Universe: Space as seen as a well-ordered whole. Cosmos expands its definition of Universe via the multiverse, parallel universes, and additional dimensions.

Evidence: The available body of facts or information indicating whether a belief or proposition is true or valid.

Evolution: The process by which different kinds of living organisms are thought to have developed and diversified from earlier forms during Earth's history.

Habitable Zone: The Habitable Zone is the distance from a star where liquid water can exist on a planet's surface.

Life: The condition that distinguishes animals and plants from other matter, including a capacity for growth, reproduction, functional activity, and continual change preceding death.

M-Dwarf (Red Dwarf): A dwarf star, ranging in mass from one-tenth to one-half the mass of the Sun, whose relatively cool surface temperature makes it appear red-orange in color.

Natural Selection: The process whereby organisms better adapted to their environment tend to survive and produce more offspring. The main process that brings about evolution.

Planetary System: A star or multiple stars that has orbiting planets. This is not the same as a 'star system', which is a star or multiple stars that is not referenced in the context of planets.

Precession: The movement of the axis of a spinning body around another axis due to torque (such as gravitational influence) acting to change the direction of the first axis.

Reason: The power of a mind to think, understand, and form judgments by a process of logic based on known reality.

Science: The intellectual and practical activity encompassing the systematic study of the structure and behavior of the physical and natural world through observation and experiment.

Scientific Method: A method of procedure that has characterized natural science since the 17th century, consisting in systematic observation, measurement, and experiment, and the formulation, testing, and modification of hypotheses.

Scientific Theory: A well-substantiated explanation of some aspect of the natural world that is acquired through the scientific method, repeatedly tested and confirmed, and carries predictive power.

Super-Earth: A super-Earth is a planet that is larger than Earth, but not so large that it has the properties of a gas giant.

Tidal Locking: Where an object's orbital period matches its rotational period. An example is Earth's moon always facing the same way toward Earth.

Mathew C. Anderson

CHAPTER 1: SETTING THE STAGE FOR LIFE AND CIVILIZATION

"The surface of the Earth is the shore of the cosmic ocean. From it we have learned most of what we know. Recently, we have waded a little out to sea, enough to dampen our toes or, at most, wet our ankles. The water seems inviting. The ocean calls."
- Carl Sagan

In *Star Trek: The Next Generation*, an entity called "Q" is teaching Captain Picard a lesson.[1] Q transports the captain back in time to a primordial hellish Earth. As volcanoes spew lava all around them with a bleak sky overhead, Q kneels down and dips his hand in a pond of a thick, oily substance beside a massive rock outcropping.

Q says to Picard, "See this? This is you. I'm serious! Right here, life is about to form on this planet for the very first time. A group of amino acids are about to combine to form the first protein." Picard squints down as if he can actually discern this activity.

Q continues, "Strange, isn't it? Everything you know, your entire civilization, it all begins right here in this little pond of goo."

The Universe had no obligation to create anything at all, yet life managed to make a grand entrance on at least one tiny pale blue dot in a truly spectacular way. Out of all the potentially habitable planets, it could be that 99.99% of them only ever produce single-celled organisms and nothing more complex. Earth may be the result of a grand natural experiment tucked away in a quiet suburban corner of the galaxy. We should feel lucky to have won the cosmic lottery! Let's explore our special home in the Universe.

A PARADISE ON EARTH

A sunny weekend has arrived, so you take your family on a road trip to some distant countryside cabin. The breeze whips through your hair, and a warmth saturates your skin from the Sun's rays that pierce the partly cloudy sky. The road follows a stream as your family glimpses a deer licking up water along the stream's edge. A fish jumps out of the water briefly. A side road through a dense forest takes you up a winding climb for another hour before you reach the cabin, nestled along a pristine glacial lake. As you get out of the car, you can hear birds singing in the trees and the croaking of frogs at the lake's edge.

What makes this scene so beautiful? It appeals to us because the planet is not only suited well for life, but life is suited well for the planet. That is to say that over the billions of years of evolutionary history, both Earth and life have changed to become more supportive of and compatible with each other. A blue sky and crystal clear lake surrounded by forests is beautiful to us because we have evolved to view it that way. It is now instinctual for most people to see beauty in nature. To be in awe of nature leads humans to treat nature with respect, which helps with our survival.

The Anthropic Principle

The relationship of Earth and its life may apply to other planets. There may be life in every system we point our telescopes toward. The pessimistic view, on the other hand, is that Earth is the only planet able to support life in the entire Universe. While we have limited knowledge of what conditions life can adapt to, we do know that in our own solar system, Earth is exceptional. Our planet certainly appears fine-tuned for life when compared to every other planet, and scientists have debated how this came to be.

One idea for why life exists at all is the anthropic principle, proposed by Brandon Carter in 1974.[2] The principle states that because life requires a very specific set of laws and conditions for it to appear, the Universe to which life finds itself must by default be compatible with supporting life, otherwise life would not have arisen in the first place. A stronger view of the principle states that only life bearing universes would ever appear anyway, and there is no alternative for life to be selected for or against.

Others use the anthropic principle to suggest the Universe was tuned in advance to suit life's requirements. Douglas Adams, an English writer, famed for *The Hitchhiker's Guide to the Galaxy*, illustrates the problem of this latest idea in this quoted passage:

"Imagine a puddle waking up one morning and thinking, 'This is an interesting world I find myself in, an interesting hole I find myself in, fits me rather neatly, doesn't it? In fact, it fits me staggeringly well! It must have been made to have me in it!' This

is such a powerful idea that as the Sun rises in the sky and the air heats up and as, gradually, the puddle gets smaller and smaller, it's still frantically hanging on to the notion that everything's going to be alright, because this world was meant to have him in it, was built to have him in it; so the moment he disappears catches him rather by surprise. I think this may be something we need to be on the watch out for."- Douglas Adams

Earth may also apply to the anthropic principle in its natural fit for life. The conclusion then is that there *should* be other Earth-like worlds in the Universe, even though we have yet to discover, or at least confirm, any of them.

Although the ever increasing count of planets in other star systems indicate that our being exceptional is highly unlikely, it is still *possible* that Earth is the only place in the Universe to support life. This possibility would suggest that the anthropic principle is incorrect, and Earth and its abundance of life is a fluke. That we have detected neither lower life forms nor more evolved life forms on other worlds can also be viewed as supporting this theory. What an epic and humbling lottery we have won indeed if our planet is the only place with life in the Universe!

AN EVOLVING PLANET

Earth was created in six days, and on the seventh day God became pooped and needed to rest... wait, that's not the right story. Earth actually formed 4.54 billion years ago when the solar system coalesced from a collapsing cloud of gas and dust.

A Shooting Gallery of Billiards

For the first few million years after its birth, our solar system was a chaotic place. Collisions of objects was a non-stop affair. Just as Earth's molten surface was solidifying, a Mars-sized object collided with our planet, and some of the debris coalesced to form the moon. The remaining debris eventually fell back to the planet.

Millions of years later there was another series of impact events known as the Late Heavy Bombardment (LHB). This second wave is thought to have been caused by a shift in the orbits of the two gas giants, Jupiter and Saturn. As these planets migrated, at times toward the Sun and at times away from it, they knocked around the smaller inner planets so much that any primordial planets within the orbit of Mercury were sent careening into the Sun, or ejected out of the solar system entirely. If this had not happened, the solar system could very well have more than a dozen planets, including several larger versions of Earth-like worlds called super-Earths.

The chaotic time of the LHB helped Earth to develop its current life-sustaining conditions. The rain of comets and meteors, rich with water, vaporized upon impact, depositing on our planet complex organic compounds. If the Late Heavy Bombardment had not occurred, or came much later, Earth might only have a fraction of the water it does now. One isolated ocean or an endless landscape of large lakes may have formed instead.

Earth's Composition

From the core of our planet to the atmosphere surrounding it, each layer plays a critical role in creating and maintaining conditions suitable for life.[3]

After the Mars-sized body hit Earth, the heaviest of materials were absorbed into the deep bowels of the planet, including precious metals like silver, gold and platinum. Nearly all of these elements sank right to Earth's core, thousands of kilometers below the surface (and completely out of reach of any possible mining operation). The remaining elements formed the outer core, mantle and crust. If life was ever to emerge on Earth, these lighter elements would be required.

Earth later obtained additional heavy elements from the bombardment of comets and asteroids. Earth's veneer of asteroid-deposited heavy metals would later be crucial for powering human civilization. Most of the electronic devices we use today have small amounts of a group of these metals called rare earth elements (REEs). There are more than a dozen rare earth elements mixed throughout the planet's crust, such as terbium, used in televisions, neodymium, used in hybrid automobiles, and thulium, used in high-efficiency lasers. Interestingly, many REEs are actually quite abundant. What gives them their name is the difficulty in separating the metals from other surrounding metals.

A Restless Planet

Because the temperature of the crust is so much cooler than at the core, the inner material convects. Convection is the process that heats your food evenly in the oven. The coils at the bottom heat the air, causing the air to rise and then cool at the top, where it falls down the sides and back to the coils. Convection in the Earth moves material from the crust down into the mantle, and eventually back up again via volcanoes. This movement also gives rise to the planetary magnetic field.

Just like how we breathe in oxygen and exhale carbon dioxide, Earth also has a natural capacity to absorb and release gases. Carbon dioxide is a powerful greenhouse gas that would cause Earth to overheat if there was no way to regulate how much CO_2 is in the atmosphere. The planet absorbs carbon dioxide out of the atmosphere when rocks and plant growth settle on the ocean bottom, where the carbon dioxide becomes baked into the crust. The heavier oceanic basalt crust is eventually pulled under the continents and into the mantle. The material melts and safely releases the stored carbon dioxide under kilometers of molten rock.

The process of carbon capture takes a minimum of hundreds of years to occur, so the current carbon dioxide in the atmosphere will likely remain for the rest of our civilization's near-term existence, unless we can artificially pull that carbon dioxide back out of the atmosphere through more expedient (and safe) means.

Earth is always changing. The planet's plates are constantly in a state of movement. While a few centimeters per year might not seem like a lot, over 100 million years this will shift a continent

3,000 kilometers! Pent up forces over decades will cause a sudden release of energy, creating earthquakes and the deformation of nearby land areas. As the continents push against each other, they will build up huge mountain ranges like the beautiful Rocky Mountains, islands like the Hawaiian Islands, and meandering rivers like the Nile River Delta and the great Amazon.

In addition to the continents, the oceans help the Earth to actively regulate itself. A significant amount of water is being warmed in some areas and cooled in others, constantly generating mighty storms. These storms begin as small, low-pressure systems in the warmer latitudes, quickly grow and migrate to cooler areas, where they eventually reach land and cause storm surges. Regular El Niño and La Niña events over the oceans affect entire hemispheres with altered wind and rainfall patterns, resulting in severe droughts like those in California, or mudslides in Leyte Island, Philippines, from too much rainfall in a given season.

Earth is also constantly moving through space, revolving around the Sun once per year and once around the galaxy every several million years. The planet also shifts on its axis slightly over thousands of years, which causes long-term seasonal changes that in turn affect the survival of life. The shifts in axis can also disrupt civilizations with drought, as was the case with the Mayans, or provide a burst of sustenance with decades of plentiful rainfall.

Shifts in axis can also cause or exacerbate ice ages. During the last ice age, sea levels dropped by dozens of meters as oceans became completely covered in surface ice. This resulted in a land bridge between Asia and America that our ancestors discovered just as they were entering Siberia. Bands of people slowly migrated from Asia to America over the land bridge for thousands of years before the bridge submerged again in newly rising seas.

EARTH'S HABITABILITY

Water is the first thing to consider when gauging a planet's habitability. On Earth, wherever water is found, life is also found. Expansive oceans help to regulate atmospheric temperatures and ensure a robust hydrologic cycle. There is not too much water though that would result in a water world (a world covered 100% by a single ocean). Water covers 71% of Earth's surface, leaving enough land area available for life to evolve into more complex forms, including a form that would one day end up building a technological civilization.

Earth's atmosphere also plays a major role in making the planet hospitable to life. The atmosphere is composed of 78% Nitrogen, 21% Oxygen, and other trace gases. Oxygen levels in the atmosphere have fluctuated in the past, from trace amounts at nearly 0% to as high as 35%. Oxygen also made – and maintains – the ozone layer, which is critical to shielding life from harmful UV radiation from the Sun. The rays that do get through can cause a severe sunburn, or worse, a facial wrinkle.

Gravity is another habitability requirement that keeps everything together. Gravity keeps our feet on the ground as the planet spins at 1,670 kilometers per hour. Earth would have to spin about ten times faster for objects to be ejected into orbit. Humans could technically survive without gravity, as we do on the international space station, but there would be serious side effects. The constant downward stress exerted on our bones keeps them from losing density. Intracranial pressure of the head is also kept in check, preventing long-term eye damage. Everything from eating, digestion, and excretion are assisted by gravity.

Earth's moon is one of the largest in the solar system. Having a large moon helps our planet to maintain its tilt toward the Sun. Earth's tilt is currently at 23.4 degrees, resulting in relatively mild changes in the seasons. Without a large enough moon to stabilize the tilt, climate shifts would cause the most severe of ice ages to occur every few thousand years, instead of every few hundred thousand years. Each ice age would be more extreme as well. Warming trends would have a similarly exaggerated effect.

The rotational rate of the Earth is important in determining surface wind speeds, and to some extent assists the oceans in maintaining an overall mild variance in global temperature. Our planet today has a pleasant average wind speed of about 10 kilometers per hour. This is fast enough to scatter seeds and other particles necessary for reproduction, but it is not so fast that trees are uprooted. When Earth was less than a billion years old and a day was just 12 hours long, winds were so extreme that standing upright would have been difficult, if not impossible.

The properties of Earth's inner material may be the most important part of the planet. Earth's active interior creates a fluidic dynamo – the physical motion of material – in the planet's outer core, which generates a globally encompassing magnetic field. The field not only protects life like the atmospheric ozone layer does, but it also keeps the solar wind at bay, thus preventing the atmosphere from being eroded. Without an atmosphere, the oceans would quickly sublimate into space.

USABLE REAL ESTATE

Earth may be a life-rich planet, but it is still carefully balanced on a knife's edge in terms of its ability to support life. In fact, the majority of its surface is uninhabitable. Getting a tan on the beaches of Bora Bora or El Nido may make us feel like we live in a global paradise, but a quick trip to chilly Antarctica, or the driest non-polar place on the planet, the Atacama Desert, we would quickly realize that not all of this world is suited to humans.

As the saying goes in the real estate business, everything is location, location, location. Just 15% of Earth's land is worth occupying, whether to extract resources or to settle. Very few would build a cabin in the middle of the Atacama Desert, or on the ice shelves of Antarctica.

While most of the Earth is not fit for human settlement, the remaining total area that is fit for our use is still generous. Even though human population is now more than 7 billion persons, there is enough area to build (though not necessarily sustain) an advanced technological civilization for perhaps as many as 15-20 billion. Still, we are ruining much of the available land, and may soon reach the point where it is insufficient not because of overcrowding, but because of mismanagement.

Regardless of a planet's size, overcrowding and resource management will likely be a consideration for any advanced civilization. Even the largest of super-Earth planets, with two times or more surface area than Earth has, could fill up with walking, talking beings at some point. By comparison, if your current 500-square meter home is full of stuff and you move into a new 1,000-square meter home, you will probably want to buy that bigger bed, larger dining table, have another child, and so on.

With so little of Earth's land being useful to humans, it doesn't help that the majority of the surface of our planet is covered by water we cannot drink. The salinity of seawater makes it unusable for crop irrigation and human consumption; in fact, less than 1% of all water on the planet can be used for these purposes. If you could gather the entire quantity of freshwater both on Earth's surface and any underground, the volume would amount to a sphere with a diameter of 273 kilometers, or the distance you would travel driving at 92 kilometers per hour for three hours.[5]

EARTH'S FUTURE

Earth has entered the 'Anthropocene', a geologically distinct epoch where human activity has forever made a mark in the geological record. Future alien archaeologists will be able to find this mark, should they ever visit and we are long gone. Although our planet is always changing and entering new epochs (usually over millions of years for each epoch), and will continue to do so regardless of human activities, eventually it will change permanently for the worse. We can observe other planetary systems to get a peek at what is in store.[5]

Here is how it is going to play out for Earth in the distant future with an increasingly elderly Sun baking its surface: during the next four billion years the Sun's temperature is going to continue to slowly increase, rising by another 10% or so. That doesn't sound so bad, until we realize that Earth is already on the inner edge of the solar system's habitable zone. As the Sun's temperature increases, this zone will move further out. While Earth will not become the next Venus, with surface temperatures that can melt lead, the oceans will still boil off. Earth's surface will be baked sterile. The upper layers of the atmosphere will be carried away by the solar wind, and any remaining water vapor will seep into space. Eventually, our planet will lose its entire atmosphere. One can take heart, though, in knowing that life has at least a few hundred million years before the Sun will start to roast us.

THE PERFECT EARTH

Earth is overall a great place for life. At first glance it may seem like all of the knobs are set perfectly to meet our needs -- water to quench our thirst, food to fuel our bodies, and oxygen to breath are all seemingly available in abundance. Humans have settled everywhere there is land and these three needs are met, and flourished, so what's there not to like?

The knobs of perfection are in fact quite a bit off from what planetary scientists can envision, as well as by anyone who takes a moment to think about places on Earth that are not habitable. What we know about the limits of chemistry, geology, and other sciences would suggest that there ought to be planets out there even more suited to life than Earth. If you had control of the planetary construction dials, what sort of improvements would you make to our planet in order to create a more habitable place for humans?

With more than a thousand confirmed planets in the galaxy so far, scientists are beginning to doubt that Earth is the best real estate in town. Our planet is certainly good, great, even fantastic for life, but not perfect. The Earth Similarity Index is a recently established measure of habitability. Hundreds of factors are taken into consideration, but most important are a planet's size, interior composition, atmospheric composition, presence of a magnetic field, and presence of liquid water. Earth scores a 1 on a scale of 0 to 1. Mercury is 0.86, and the moon is 0.56. While we haven't yet discovered any planets higher than 1, the scale does allow for this possibility with some of the suggested changes below.

Clean water and safe food are the first two things that come to my mind when thinking about what a perfect Earth would need in greater abundance. It is ironic that our planet has more water than land surface, and yet thousands of children die every single day from a lack of safe drinking water. Even when the water is not contaminated, if it is salty like that of Earth's oceans, drinking it will only cause one to dehydrate faster. Hundreds more children are poisoned from wild or improperly prepared foods.

A higher level of oxygen in the atmosphere would help in fueling our bodies for very energy-intensive tasks. With more oxygen in the air, lung capacity would increase, requiring a smaller set of lungs. The reduction in lung size would free up the body's resources to focus on other operations, perhaps increasing the brain's efficiency, enhancing intelligence. It took Earth a billion years to build up enough oxygen to fuel multicellular life, so it's clearly a key component in making complex life happen.

More oxygen would come at a cost, however. There would be an increased chance of wildfires. The doubling of oxygen to about 40% could cause wildfires to burn out of control even during a heavy rainstorm. This amount of oxygen in the atmosphere was last seen about 300 million years ago. Since then, levels tapered off to the current 21%. Any lower than 15% and extinction events could occur, alongside an onset of sudden evolutionary changes in species in a struggle to adapt.

We might also think that a perfect Earth would have no natural disasters, but as described above with atmosphere regulating plate tectonics, a planet cannot be unmoving and expect to support something as dynamic as life itself.

Understanding what is life, what makes our planet capable of supporting it, as well as life's limitations, will help us in the distant future when humanity looks toward space for another place to call home. Until we find such a world that is worthy of being called Earth 2, the worst thing we can do is shrug off the responsibility of taking care of our one and only home. That process of understanding begins with the origin of life itself.[7]

CHAPTER 2: EVOLUTION AND THE BUILDING BLOCKS OF LIFE

"The theory of evolution by cumulative natural selection is the only theory we know of that is in principle capable of explaining the existence of organized complexity."
- Richard Dawkins

THE GREATEST EVENT IN HISTORY

A miraculous event occurred four billion years ago. On a watery world safely tucked away between two arms in a spiral galaxy, organic compounds arranged in such a way as to be able to self-replicate. Once replication was in place, incremental changes started to occur in future copies. These changes arose from copy errors, damage, or stress which altered the original set of information, occasionally producing new helpful variations that were passed on to the next generation. This process, called evolution by natural selection, started a chain of events that led to the existence of thinking beings that can reflect on that original spark of life.

Let us examine how life got started on Earth, and how it has flourished over time to become as beautifully diverse as we see it today. Understanding the life on our planet will tell us a lot about the chances for it to occur elsewhere in the Universe.

Place the chemistry of life and the process of evolution by natural selection in a stable but slowly changing environment (Earth), stir in a whole lot of time (billions of years), put it on the fire of continuous energy (the Sun), and *voilà*, you get millions of species coexisting, as they do today.

THE FOUNDATION OF LIFE

Life: *"The condition that distinguishes animals and plants from inorganic matter, including the capacity for growth, reproduction, functional activity, and continual change preceding death."*
– Oxford English Dictionary

The Periodic Soup

Regardless of which nebula or gas cloud in deep space we point our telescopes toward, we discover the basic building blocks of life – the elements – being formed therein. The laws of chemistry and physics dictate how these building blocks come together in predictable and consistent ways.

Currently there are 118 elements on the periodic table. Elements either repel, bond, or remain neutral to other elements. If the atom of one element bonds to an atom of another element, a compound is formed. Compounds are two or more atoms that bond together, and they are called molecules when they have a neutral charge. Complex molecules, such as amino acids and proteins, allow for life to form out of the elements.

Because of its abundance and unique ability to easily bond with other elements, carbon is the element best suited for the formation of complex molecules advantageous for life. Another advantage of carbon is that its bonds split and recombine easily, but not so easily that they cannot hold long enough for life to develop. Carbon also bonds easily with other elements that are very abundant in nature, such as hydrogen, oxygen, and nitrogen.

There are only a few other elements around which life-yielding molecules could possibly form. Silicon is considered the next best candidate, but it is still far behind carbon. Silicon requires more energy to combine with other elements, and has less combining opportunities. The more energy required, the less likely life will be able to take hold. For example, a lightning strike may be needed to form a silicon bond of sufficient complexity, versus the Sun's natural light striking the ground's carbon atoms, forming bonds.

Water – two hydrogen atoms and one oxygen atom (H_2O) – is the molecule most valuable to life.

Liquid H_2O is the best solvent for chemical interactions, and it is widely agreed that life started as a chemical reaction in water. The oceans circulate organic molecules around the globe, where they can lodge into more secure environments like a small bay. As the moon's tidal forces act on the bay by repeatedly raising and lowering the water level, the chemical soup begins to form layers. Heating and cooling from the planet's day/night cycle then cooks the chemicals together into increasingly complex forms. Eventually something will change in the mix and combine in such a way as to spawn self-replicating molecules.

Life has a propensity to evolve every chance it can get, and it is generally very successful, barring an asteroid impact, a massive volcanic eruption, or the meddling of other life. It just needs that initial chemical kick-start and it's off to the races.

Life's Scaffolding – RNA and DNA

Many scientists think that RNA (ribonucleic acid) was the first self-replicating molecule. RNA is a long strand of molecules, made up of sugars and phosphate bases (nucleotides). RNA and DNA (deoxyribonucleic acid), an even more complex strand of molecules, exists in every living cell and contains the genetic information to be passed down to offspring.

A Synthetic RNA and DNA – XNA

A few alternatives to RNA and DNA have been created in laboratories. The leading one is xeno nucleic acid (XNA), which has almost all of the same components as RNA and DNA and is able to store information and self-replicate.

Genetic engineers use XNA to create more energy-efficient chemical reactions in cells than they can create with RNA or DNA. They are working to find a way to use XNA to target cancer cells and fix damaged RNA and DNA. Critics worry that XNA could infiltrate the ecosystem and contaminate it, but genetic engineers claim that the risk is negligible.

No matter how intricate life's copying processes get, it's not going to allow us to fly without wings or shoot fire out of our eyes, as you may be led to believe by fanciful movies like *Superman* and *X-Men*. The physical limits of the formation of molecular structures give us one reason that alien life is likely to be more similar to earthly life than we might first imagine.

THE BEGINNINGS OF LIFE ON EARTH

Environmental conditions had to be stable for a long enough period of time so that the initial molecules capable of replication could form in sufficient quantities. The process could possibly have happened many times over before it took hold; it may be that each time it got started, the beginnings of life were destroyed or disturbed sufficiently to abort the process.

The atmosphere on early Earth was made up of hydrogen sulfide, methane, and high levels of carbon dioxide. This mix was beneficial for the original single-celled organisms, and it caused their population to explode. These organisms produced an abundance of oxygen as a waste byproduct, and for more than 2.3 billion years this oxygen saturated the oceans...the stage was now set for complex life to evolve and thrive.

Just a few hundred million years after oxygen saturated the oceans and began filling the atmosphere, the first eukaryotic cells evolved. These cells have a more complex internal structure than their prokaryotic ancestors, as they contain a membrane-bound nucleus and internal organelles, such as mitochondria that produce energy for the cell. Before this period, all cells were prokaryotic and could not combine to form multicellular organisms. Once eukaryotes evolved and made use of the abundance of oxygen, complex multicellular organisms soon followed. The first species to reproduce via sex, as opposed to asexual reproduction, also occurred around this time.

Life can now get interesting! In fact, all life on Earth today has evolved from a long chain of reproduction that started with those very first eukaryote cells.

While oxygen is one element that is needed for complex life, it is not the only one; life forms must ingest other elements as nutrients. At the same time that some life forms gained the ability to use oxygen, others lost the ability to carry out photosynthesis, the process that had supplied these other elements. Life forms that thrived on oxygen began to ingest the single-celled organisms, which still performed photosynthesis, and in this way they fed themselves the elements they needed. The food chain we partake in today began at this oxygenation event.

Over the next 1.5 billion years, oxygen continued to be produced as a byproduct of photosynthetic life forms, and over time enough oxygen existed to form the ozone layer in the atmosphere. Ozone, O_3, shields Earth's surface from the Sun's deadliest radiation. Previously, life was confined to the oceans where water provided a sufficient shield. This protective ozone layer allowed even more complex organisms to develop, and for life to move out of the water and onto land.

Starting about 540 million years ago, millions of species appeared in a relatively short time. Scientists are not in agreement about how long this period lasted; opinions range from a scant 5 million to a much more plausible 80 million years. This upsurge in diverse life forms is called the Cambrian explosion, and was the precursor for the age of the dinosaurs. Many of the key organs and other features future land animals would acquire, such as a heart, brain, and backbone, also evolved during this time period.

LIFE'S ABILITY TO EVOLVE

In terms of geologic history, Earth's changes are slow and steady, apart from infrequent cataclysmic disruptions, as in the asteroid bombardment that may have caused the dinosaurs to go extinct (clearing the way for mammals to evolve). Seasons change every few months, and the more dramatic ice ages occur as frequently as every ten thousand years, and as infrequently as every hundred thousand years or more. It's a slow enough rate of change to allow life to adapt.

Life alters its own environment, which can have the result of allowing new life to form. As explained above, life that put oxygen into the atmosphere (plants) allowed oxygen-breathing organisms to evolve (animals). If carbon dioxide levels in the atmosphere rise, more plants grow, which helps to remove any excess CO_2. None of this is through foresight; nature does not plan for the future.

Whatever proves most beneficial, and is physically and chemically possible, will generally occur.

Life has been found in some of the most inhospitable places on Earth. Deep in the oceans where light cannot reach, life forms called extremophiles cluster around hydrothermal vents that spew out 260°C sulfuric gases, which would be poisonous to other life forms. These extremophiles have traits which make them well adapted to the harsh environment. Life is also found high in the stratosphere where it's cold, extremely dry, and abounding in solar radiation.

And yet as we see with Mercury, Venus, Mars, and every other body in our solar system, life cannot form in just any environment. If life could appear anywhere, the Universe should be a very busy place!

Richard Dawkins – Mount Improbable

Richard Dawkins is an evolutionary biologist and humanist. Dawkins has written many books on evolution and participated in numerous debates on the subject.[1] He is also an ethologist (the study of animal behavior). Dawkins started out studying biology and teaching at the University of Oxford as Professor of Public Understanding of Science. Recently, he has been focusing on the public's awareness of critical thinking skills, or the "raising of consciousness" as he calls it. Because of his experience in biology, Dawkins is front and center when it comes to evolution.

Three of Dawkins' very popular books related to evolution are *The Selfish Gene*, *The Extended Phenotype*, and *Climbing Mount Improbable*. In *The Selfish Gene*, and partially in *The Extended Phenotype*, Dawkins' talks about how an organism is only the tool genes use to ensure their survival. In *The Extended Phenotype*, Dawkins' goes further and points out the effect an organism has on its surrounding environment, based on the behavior of that organism guided by its genes. In both of these books, the genes are the master commanders, not the host organism that does the gene's naturally guided bidding.

Dawkins' third book, *Climbing Mount Improbable*, explains how evolution occurs through natural selection.[2] Evolution is not a series of direct improvements in a species. Evolution is about change over time, usually guided by environmental pressures that force a species to adapt to its surroundings. Changes often first appear as errors in the genetic replication process, known as mutations. Changes that are harmful and do not improve chances of survival will less likely be carried on to the next generation. Far fewer beneficial changes are going to occur than harmful ones, but they still do occur. Given enough time, these beneficial traits will naturally be selected for and spread throughout the population of a species. If a segment of that species becomes isolated for long enough, and new traits appear which favor that segment's new environment, a distinctly new species may eventually result.

Evolutionary Pressures

Life may not change from generation to generation if there are no pressures for it to change. If the environment stays exactly the same for millions of years, life in that environment may remain nearly identical for that entire time. There are organisms on the ocean floor that exhibit this lack of change. We know, though, that most environments do change, and this puts pressure on life to adapt, and the need to adapt results in changes in a species over time.

Survival of the fittest is the classic theory of how life evolves via selective pressures against other lifeforms. The weak die off and the strong remain, and they go on to reproduce offspring with those traits that made them strong. Useful traits continue to remain in a species, while others that are no longer useful will be reduced in size and function, or removed entirely.

Change in climate is another key pressure that can cause a variety of changes in life forms. For instance, 56 million years ago, during the Paleocene-Eocene Thermal Maximum, the climate was several degrees warmer than it is today. The first horses evolved then, and they were less than half their modern average size. In general, the warmer the climate, the smaller mammals tend to be. Smaller bodies are able to shed heat more quickly and thus keep cool. As the climate cooled, mammals like horses grew in size. Change in climate also alters sea levels, which can separate species. Over time, the separate group becomes an entirely new species.

Sources of food can dissipate, pressuring species to migrate to new environments. For instance, some birds today are having to fly further to find food that is no longer at the latitudes they used to hunt. The effort in flying great distances is improving their flying skills and likely causing these birds to become stronger. These birds will pass down genes that empower them to meet their needs to find food. If a key food source becomes scarce, especially for species that have less mobility than birds, extinctions may occur.[4]

Emotional stress as a selective pressure has gotten a lot more attention recently. Like with geographical barriers, stress also plays an important role in driving speciation. Stress helps us to avoid approaching danger, and it gives us energy to overcome otherwise insurmountable obstacles. Stress also acts on our genes by activating or deactivating them as a result of significant events. Activated genes are those that become chemically strengthened, causing them to be more likely to be passed on to offspring in an active state.

CLIMBING "MOUNT IMPROBABLE"

Even though evolution is based on the hard limits of chemistry and physics, the abundant variety of life we see around us is astounding. If evolution had taken another course, perhaps because life found itself in a different climate or other pressure, we can only imagine what sort of unique traits and creatures would have resulted. Earth has numerous examples of life taking unusual and *improbable* paths with varying results. Just think about how many varieties of birds, insects, mammals, and other classes of life there are around the world.

As much as there are constraints on life, we also know that certain features and functions seem to be an absolutely necessity to survive in most environments.[3] Life has numerous examples of independently produced features. Convergent evolution is the same feature evolving in more than one species when those species have no common ancestor to share genetic information about reproducing said feature. A few examples of convergent evolution include bird wings, fish fins, arms and legs, powerful foreclaws, long and sticky tongues, and the eye.

The Eye

The earliest version of the eye is thought to have been a grouping of light-sensitive cells on the dorsal surface of some fish. There was no retina, cornea, or lens. As these fish swam in shallow waters, light from above would hit the cells, providing the fish with an ability to sense depth. Over time these light-sensitive areas evolved, changing to allow fish to sense the shadows created by the predators passing overhead.

Over more time, a mutation in the fish's genes created an indentation in the light-sensing area. The mutation benefited the fish by allowing the detection of light from more than one direction. This concavity was the first evolutionary move towards an eyeball. Eventually a pinhole formed on the indentation – an early pupil. Some species would also develop a thin sheath over the pinhole that helped to protect it – this sheath evolved into the cornea, the surface of the front of the eye. Versions of this sheath helped focus light coming in through the pinhole; eventually a lens evolved and the eye of today's abilities appeared.

Some fish evolved limbs with which to climb onto land. Soon after, eyes evolved to be located more toward the front of the head. This frontal view greatly improved the ability to detect the next meal, shifting the role of the fish from prey to predator. For every generation that had slightly more forward-facing eyes than its parents, the ability to capture prey increased.

Every stage of evolution the eye has gone through can still be found in creatures today. For example, the eyes of vipers and some pythons are simply shallow holes with light-sensing cells. Giant clams have a pinhole that allows them to navigate the complex and vast structures of coral reefs. The Proteus anguinus salamander, having migrated to dark caves, no longer needs its eyesight, so its eyes have regressed. The larvae start off with eyes, but then they atrophy because they are a long unused organ of the species.

Some consider the eye to be a perfect organ. After all, we can see incredible amounts of detail in millions of shades of color. For some species, eyes work extremely well under water, and for others, eyes can spot prey in the pitch black of night. Indeed, the eye is an amazing organ, though it is far from perfect. There are blind spots, perception inconsistencies, and age-related disease of the eye. Statistically, 17% of the world's population is blind to some degree, and about 1% of children are born blind. Every year, more than a million children are born never to see a colorful flower or the smiling faces of their parents.

Insects

Insects provide an excellent example of how the environment shapes the evolution of a species. As mentioned previously, insects absorb oxygen through their exoskeletons, via little holes, called spiracles. The greater the surface area of an insect's body, the more oxygen it can absorb. The more oxygen in the environment, the larger insects become – though they would never become the size of a human due to fundamental physical limitations. During the Carboniferous geological period more than 300 million years ago, atmospheric oxygen levels reached 35%. This may have sparked a period of insect gigantism, where insects were many times the size we see them today.

From Water to Land

One of the most important events in the animal kingdom's evolutionary history, that would one day lead to human beings, was the migration of creatures from water to land. Some species evolved to be able to rest partly out of the water along shallow shorelines. The fins of many of these shallow water creatures, over many generations, evolved to form stiffer and more complex hind limbs that could push the creature out onto dry land. The first fish/amphibian hybrid was known as the Tiktaalik. This was also the first time that primitive lungs evolved.

While the exact reason *Tiktaalik* made the transition onto land is unknown, it's likely to have been the need to escape predators in the water. Since there were no other creatures on land at the time, retreating to land to rest was a perfect survival tactic.

EVOLUTION'S PROCESSES

As stated, the process of evolution is simply biological change over time. Change occurs over the course of many generations of a species. Each generation will be ever so slightly different from the previous generation, and from its offspring. Changes occur for a variety of reasons, from environmental to societal, from accidental to man-made.

Hundreds or more generations are required for significant changes to manifest in a species, so such changes can be difficult to identify. Let's envisage a human lifetime displayed in a picture book, with one image per day on a page. On average, there would be more than 25,000 pages. Now let's remove every page but the first and the last. The first page would picture a just fertilized egg, while the last would picture an elderly person on their last breath. It might initially be difficult to understand how the two images were related, until we take into account the role of change over time.

Unguided Man-made Evolution

The scale of time over which evolution occurs should not affect our understanding of the process, but even so, it is human nature to doubt that which we cannot observe. Yet there are species that have evolved quickly enough for us to better grasp the concept. Those creatures are right in front of our eyes and licking our noses. I'm referring to our beloved pets, of course – namely the dog.

Dogs are descendants of wolves. Humans and wolves have had a relationship going back for tens of thousands of years. Originally, wolves would appear on the outskirts of human habitats, in search of food. Humans would leave food out, much like how we may, perhaps unwisely, leave food out around a campfire for hungry creatures in the night. This was more than an act of kindness, though, as wolves kept away more dangerous creatures, such as poisonous snakes. Over time, the wolves became comfortable with approaching the humans, and the relationship between humans and wolves grew. As more sophisticated settlements with greater populations sprouted up, wolves would venture into villages and towns to look for food. Wolves eventually became a regular part of life in the settlements.

Once domestication began, the wolves' physical features underwent evolutionary changes. Where high testosterone once helped wolves protect themselves, such high levels were no longer needed once humans were keeping the wolves safer by caring for them. Lower testosterone levels reduced rigid facial features, making wolves appear softer and less aggressive. (High testosterone gives chimpanzees their enlarged brows and makes them some of the most territorial of animals.) Humans would naturally favor wolf cubs that had less intimidating features; being favored by humans helped wolves survive, and so these physical changes accelerated.

Eventually, wolves evolved into dogs. Neither wolf nor human was aware of the process underway at the time, and most still are not aware of this transformation still occurring today.

So why are there still wolves then? Wolf populations span entire continents, so naturally some lived in the wilderness on their own. Group separation is one way evolution forms new species. The wolves of today are so different from wolves of millennia ago that they are considered two distinct species. By extension, humans did not evolve from the apes that are around today, we just share a common ancestor.

Guided Man-made Evolution

Another way evolution sculpts species is through reproductive constraints. For instance, a Great Dane and a Chihuahua cannot naturally copulate because of the great difference in their sizes. In the case of dogs, humans guided the process with selective breeding. Today there are entire factories for breeding dogs, all of which are still the same species split off from a wolf ancestor many generations ago.

Because of human guidance of genetic selection, we have seen more than a tenfold increase in crop yields. Increases did not just come from improved irrigation and disease management. They also came from humans selecting the healthiest plants and taking their seeds to plant elsewhere. Growth density, plant height, ability to weather storms, and many other factors were considered in plant selection. If a crop had 20% blight, it would not be selected, possibly causing that strain to disappear from the genetic pool entirely.

There is no mechanism of a plant (or any other form of life, humans included) that says, "Stop! I'm not changing anymore. I like the way I am now!" If change continues over a sufficient number of generations, especially if a plant becomes isolated from its native environment, future generations will be unable to germinate with the source plant, and voilà, you get a new species.

Genetic selection applies to crops low to the ground, as well as those high up in the trees. The yellow bananas found in supermarkets today came about over hundreds of years of selective human cultivation from the plantain. The plantain came from western Africa and is about a third the size of yellow bananas, darker in color, has a tougher skin, and tastes starchy. The plantain itself has been selectively bred over thousands of years. Its earliest form was small and extremely starchy, and the plantain tree itself was more vulnerable to changes in climate.

Another example of human guided evolution would be with corn, which is at the very foundation of our food supply. While corn is native to North America, it has since been grown all over the world. The vast geographic dispersal has slowly diversified corn's appearance. While all varieties are still of the same species, corn today looks nothing like when Indians began harvesting it thousands of years ago. The original plant more closely resembled wheat. It was thin and short with a few tiny seeds along the branches. Steadily over hundreds of years, the tastier varieties with larger kernels were cultivated.

What are the consequences of man-guided evolution? Man has cultivated corn to such heights, the heavy cobs tended to drag down the plant's branches, and would topple over a plant on a gusty afternoon. Left to grow on its own, corn would quickly regress back to a smaller size after just a few generations, or even go extinct. If humans disappeared tomorrow, so would a majority of our crops within a matter of months, not to mention many other plants and domesticated animals.

Breeding a very specific crop can result in less genetic diversity. When reproduction occurs naturally, there is not as stringent a filter on which genes are passed down to future generations. Species are partly protected from going extinct by having diversity in their genetic makeup. A large gene pool will help a species survive epidemics, for instance. Life on Earth has a remarkable propensity to maintain itself by diversifying. This mechanism can be viewed similarly as an investor diversifying his portfolio to prevent catastrophic financial losses.

Horizontal Evolution

A more recent theory of evolution suggests genetic traits are not just passed on through asexual or sexual reproduction, or guided by another creature, but also through a process that transfers genes directly between species, called horizontal gene transfer. Genes are borrowed from another nearby organism. The process is rare, but it does occur, mostly to microorganisms like bacteria and viruses, but also with plants. There is also horizontal gene transfer within the animal kingdom, with examples being rotifers and some beetles. When we create hybrids by splicing their roots and mixing them with other species, so that the new plants can then reproduce on their own with the new gene set, we achieve man-guided horizontal evolution. Horizontal gene transfer has been the source of drug-resistant strains of bacteria, as well as resistance to many pesticides.

The Ugly Side of Evolution

The food you ate yesterday probably came from a farm. That farm had both natural and chemical fertilizers, some produced from by-products animals previously had excreted. Broken down by microbes and smaller life forms, those by-products eventually settled into the soil. Basic elements and nutrients were then free to disperse further into the environment, where they could be used by whatever would grow there. The cycle then repeats.

We live in a strange world – a world where life literally lives off of other life. Life feeds off of life by necessity, sometimes causing pain, suffering, and fear in higher animals. The basic need is fantastically barbaric, and yet we are all participating in the process on a daily basis. Even strict vegans are still eating life, whatever non-meat or non-dairy substitute they may choose. While feeding off other life forms seems to be necessary, did the process really have to be this way? Couldn't life have evolved a bit differently, perhaps so that we could survive and reproduce just on sunlight, chemicals, or other non-biological compounds?

Although some may find it unfortunate, the answer seems to be no. Complex life cannot survive off the energy from the Sun or solely any other inorganic matter – consuming other life forms is a natural process that has no known alternative. Anyone who would wish for an ability to survive without consuming anything that was once alive must come to terms with the fact that nature does not concern itself with moral imperatives, that nature will take the path that tends to be the shortest and least energy-intensive, and that life comes from life.

SIMULATING LIFE'S EVOLUTIONARY PROCESS

We do have one ace up our sleeve in gauging life's potential, and that lies in understanding how the laws of physics constrain the diversity of life. As described earlier, life is limited on Earth in certain areas for obvious reasons, like the environment being too cold or too dry. Applying this knowledge of life's constraints to models on powerful supercomputers, we can extrapolate what other chemical reactions, and more complicated evolutionary processes, are likely to occur when the system is given a chance to play out over millions of years.

If you want to see evolution in action on the computer screen, Conway's Game of Life allows for such observation. The game is a simple simulator that shows how blocks interact with their neighbors when following certain rules, set by the player, and how these blocks can form increasingly complicated systems that self-replicate and evolve. The variety of what is produced in a short period of time is impressive. While just a simulation, it provides an excellent, albeit simplified, illustration of the process of evolution.

RESOURCE LIMITATIONS

One reason the human population did not grow very much for millennia was that resources were limited, but once new energy sources were discovered a few centuries ago, the population exploded. Understanding how to use more complex energy sources gave us a new ability to provide not only more secure food and shelter, but also education and medical care to extend the lives of human beings.

The limitation or drastic change in resources can open up evolutionary opportunities as well. For example, after a massive asteroid hit Earth about 65 million years ago and severely altered the climate and food supply, dinosaurs went extinct, which allowed new species to propagate. Small mammals grew more dominant because there were fewer large predators to impede their evolution.

UNIQUE HUMAN CAPACITIES

The following traits may or may not be required for a species to one day be able to develop a civilization, but they certainly help maximize the potential. In combination, the traits have allowed modern humans to improve their living conditions far beyond what is possible in other species without them.

Opposable Thumbs – Manipulating the Environment

An important trait of any species that has the chance of creating a complex civilization is the ability to significantly manipulate its environment. If there is no way to create tools that lead to more sophisticated ones, it will be impossible for a technological civilization to get off the ground.

To finely construct tools, a being needs a grasping mechanism that is able not only to carry objects, but also to manipulate its parts. Fingers clasped together pushed against other fingers will do a pretty good job at basic tasks like moving material from one location to another, as will the beak of a bird. But as for crafting tools, manipulating materials and building structures, there needs to be a grasping mechanism that includes a clutching action, and that's where the opposable thumb comes in.

The opposable thumb probably originated as a birth defect. Individuals with such a beneficial trait were more readily able to survive, as well as prove useful to those around them. Offspring from these individuals had a greater chance of acquiring opposable thumbs and, in turn, surviving. Not only was the trait passed on, but it also continued to change and become more useful. Each creature that had a slightly more pronounced opposable thumb had a greater capacity for manipulating the environment than those that came before. So although a mere genetic abnormality at first, the opposable thumb became a crucial part of what made society possible.

Bipedalism – Transportation

Millions of years ago, apes used their powerful elongated hands to grab tree branches, just like many species do today. As the trees thinned from an increasingly arid environment, eventually small patches were all that was left of a once great rainforest canopy. This forced the apes to travel on the ground to look for food. Hunting food on all four limbs proved too slow and cumbersome, so over time the skeletal structure evolved and apes began to walk upright.

The ability to walk upright proved to be both a blessing and a curse. The hands were now available for a variety of purposes, including toolmaking, communication, and crafting weapons to throw at prey. Unfortunately for the females, standing upright caused the birth canal to narrow, which made giving birth very painful. The benefits must have outweighed the cost, otherwise evolutionary changes would likely not have occurred.

Walking upright provided unique potential for our species, and it could be the most important factor in the evolution of intelligence, apart from a large brain and opposable thumbs. Walking upright preceded the higher functions the brain eventually acquired. Because the hands became available for other tasks than walking, Homo erectus and then Homo sapiens were eventually forced to think in new and different ways, which pushed the brain's development.

The Human Brain – Intelligence and Forethought

Everything from ants to the great sperm whale has a brain of a size that is roughly scaled to its body weight. This is not a hard rule though, as huge dinosaurs had tiny brains. For instance, the mighty stegosaurus had a brain the size of a bent hot dog. The velociraptor had a brain the size of a plum, which is about the same size as the brain of today's common rabbit.

Intelligence level is not entirely determined by the size of the brain, though. Sperm whales have the largest brains of any animal – five times the size of humans' – yet whales do not have the ability to construct a civilization like humans have done, the lack of opposable thumbs notwithstanding.

Our ancestors maintained an intelligence level similar to other primates for millions of years, but then environmental changes forced them to use more complicated survival techniques, like that of toolmaking, which in turn required more intricate brain functions. With a changing environment and growing population, there was a need to find alternate food sources, fend off new types of predators, and adapt to more complex social elements, all of which played a role in increasing intelligence. This period in history began a runaway effect of increasing brain complexity.

With such a powerful tool as the human brain and its trillions of neural connections, it is not surprising that Homo sapiens was able to advance quickly after it began making tools, living in large groups and communicating with language. For the first time on Earth, a species had the capability to plan and invest in its future. Once it could plan – and had learned how to kill protein-rich animals for food with crude weapons and command fire to cook the meat – constant foraging became less important. Those who thought faster and in smarter ways survived more easily than those who hesitated on the hunt for their next meal, or perhaps their next mate.

Vocal Communication

Vocal cords allow humans to use a wide range of sounds that over evolutionary history came to have meaning to listeners. If humans did not have vocal cords at all (or lacked ears with which to hear them) and could only communicate nonverbally, our hands would indeed be very busy. Vocal communication freed up our hands to perform toolmaking and fight off predators more effectively. Communication improves survival; for instance, it is advantageous if a species can quickly warn its fellows of danger. Communicating with sound enables us to receive information without having to see or be near the communicator.

Relaying information and expressing ideas to others is important for a civilization to form and eventually be able to advance in technology. Communication of knowledge, passed down from generation to generation, allowed us to grow intellectually. The written word followed – again, thanks to that opposable thumb – but vocal communication is one of the most rapid (though not the most accurate or permanent) forms of relaying information.

HUMAN VESTIGES

While the human body has many useful features, it also carries around parts that are no longer needed, or at least not absolutely essential. The appendix is part of the digestive tract that, thousands of years ago, was used for the digestion of plant material. As toolmaking improved, so did the yields from hunting, so parts of the digestive tract developed further in order to digest more meat. The appendix is still around because this change in diet is relatively recent to our species; in the future, our descendants may not have an appendix at all.

Wisdom teeth are also no longer required. They were once useful in the digestion of plants; as heavy grinding was necessary to break down tough leaves before they were processed in the digestive tract. The appendix and wisdom teeth are the most well-known vestiges because they tend to cause us misery when something goes wrong with them. When someone has an infected appendix or wisdom tooth, these vestiges can be surgically removed before they cause further harm, or even death.

While human musculature is not vestigial, it has undergone considerable reduction in strength from what our ancient ancestors had. Even the might of today's greatest heavy lifters pales in comparison to that of many other species of similar size; for example, the orangutan has a grip strength four times that of a human. It is postulated that the change in humans occurred because our brains grew in size; natural selection may have reduced the bulk of our muscles to conserve energy and oxygen for our large brains. The human brain uses 20% of the calories and the oxygen required to run the body. Also, with the greater intelligence that came along with a larger brain, we needed less muscular power to dodge predators and hunt for food.

The human ear's complexity pales in comparison to that of many other animals' ears. The macaque monkey can move its ears in full rotation in order to capture sounds from all around. The best we can do is move our head and body. Humans still have many of the same ear muscles as the macaque, but they are minimally developed and non-functional.

Take a look in the mirror and carefully examine your eyes. You may notice that the inner part that rests alongside the nose has a small, fatty bulge. This bulge is a curled up bundle of inert muscular tissue that controls a nictitating (blinking) membrane, a secondary eyelid that other mammals have but which humans do not. (Cats have such a membrane.) Like wisdom teeth, our nictitating membrane muscles may eventually fade from the species, if nature does not find them useful again in the future.

WHERE WILL EVOLUTION TAKE HUMANITY?

One day in the future, it may be the fate of humanity to change to fit some new environment – one that is, sadly, not conducive to civilization. Future generations may have reduced brain size, with less potential for intelligence, in order to conserve resources for a new survival need. All of the aforementioned traits may not be required for life to exist and reproduce. The end-game for humanity may therefore not be to maintain intelligence, civilization, benevolence toward others, appreciation of beautiful music and art, and the ability to colonize outer space, but simply to survive.

This is not to say that life is meaningless. The driver of evolutionary change may be the genes, but the meaning of life and all its grandness is squarely in the hands and minds of the conscious creatures that are host to those genes. Consequently, the future of life and the ultimate fate of its spawned intelligential creatures may eventually fall out the hands of its genes, and into the hands of the conscious creatures that will manipulate those genes for their own purpose.

Mathew C. Anderson

CHAPTER 3: THE RISE OF CIVILIZATION ON EARTH

"The Earth is a very small stage in a vast cosmic arena."
– Carl Sagan

EARLY HUMAN CIVILIZATION

The Paleolithic era, or Early Stone Age, is the earliest period in which we can identify ancient humans as a distinct species from previous ancestors. During the Paleolithic era many Homo sapiens still lived alongside the Neanderthals and Denisovans, both close relatives, genetically speaking. There is even evidence of the three interbreeding. Some scientists believe that one reason the Neanderthals eventually disappeared is that the Homo sapiens population of the time was comparatively much larger, and the Neanderthals were simply absorbed into their genetic pool. Because of this cross-breeding, many of us have a little bit of Neanderthal in our DNA.

After a long and barbaric Stone Age that lasted millions of years, other ages that were relatively more civilized quickly followed.[1] The Neolithic period came at the end of the Stone Age and it marked the beginning of humans settling and cultures forming. The Neolithic Revolution was also the transition from hunting and gathering to the use of agriculture. A side effect of humans roaming less and settling more was the encouragement of cooperation and exchange of ideas. Language flourished, as well as the need for social stability and security.

As society became more complex, the demands upon our brains increased. Agriculture helped to feed an energy-hungry brain that from generation to generation evolved in complexity. The very concept of progress and advancement was also born around this time. Once the secrets to creating a stable and productive society were discovered, vast amounts of time were freed up. New leisure activities like art, music, and science developed. For the first time in all of history, a creature on the planet had the capabilities to craft its own future, and even have fun in doing so.

The Neolithic Revolution was a time when each new discovery was changing the world. The concept of schools had yet to come about, but many cultures still shared resources and processing techniques. Regional knowledge expanded and, over time, much knowledge became global. The wheel is an excellent example of an early tool that improved the efficiency of countless human pursuits. When discoveries and inventions transformed the civilizations, a new technological age was ushered in. The next to follow were the Bronze and Iron Ages.

The Bronze Age is so named because it was the era when we discovered how to make bronze alloy. Bronze is a mix of copper, tin and a variety of other metals, such as zinc and nickel. When heated and melded properly, extremely durable tools can be made. The alloy marked the first time we could build things with materials beyond the raw resources that were cut down, dug up, sculpted, etc. Bronze was also an essential component in some structures, including the first plumbing system in the Indus Valley Civilization around 2700 B.C.

After the Bronze Age, we learned how to make even stronger metals. Iron (and later its steel alloy) was first forged at the start of the Iron Age. Iron was not discovered previously because it was so difficult to mine – an even more arduous process than figuring out the right metal mix to make bronze. The process of mining was also a new concept. As mining spread from village to village, blacksmiths quickly popped up. Iron was used to create a wide variety of items, especially deadly weapons. Bronze swords, in contrast, just didn't cut it on the battlefield.

The Iron Age also marked the appearance of the first true alphabet, which came from the Greeks. Unlike Greek's precursor, the Phoenicians, letters represented both consonants and vowels, as opposed to only consonants. The Greeks were mighty traders of the Mediterranean Sea during the Iron Age. Travel and trade was a way of life. In order to keep track of everything they traded within such a diverse set of lands and associated languages, they invented an alphabetical system of writing in order to make record of transactions. Alphabets became a standard of nearly every civilization thereafter.

THE FIRST CIVILIZATIONS

Many civilizations changed the course of history for humanity. There are six great civilizations recognized as being among the first: Sumer (Mesopotamia), Ancient Egypt, Norte Chico (Mexico), Olmec (Mexico), Indus Valley (Pakistan), and China. While only a couple have survived the test of time (Egypt and China), each of them contributed important elements to the development of today's modern civilizations that are now capable of reaching for the stars.

Sumer

The first civilization to appear in southern Mesopotamia was in Sumer, which is known as the "cradle of civilization." The Sumer civilization began around 4000 B.C. and lay between the Tigris and Euphrates rivers, in today's Iraq and Kuwait. The Sumer built a wide array of cities, which were the first examples of urban life. Walls were built to keep out invaders, which worked for at least a few hundred years. Unfortunately, the main wall (about 250 kilometers in length) was not well defended, and there was nothing to stop the hordes from simply walking around the sides to pillage the towns and cities.

Sumerians invented many things, including the wheel, the plow, and glass. The first inklings of organized religion started in Sumer. One reason that the Bible can be interpreted to suggest Earth's age at a scant six thousand years is that the first drawings and writings we know of were created six thousand years before Christ, in Mesopotamia.

Ancient Egypt

Around the time that Sumer sprouted up, more than a thousand kilometers to the west, Ancient Egypt flourished along the Nile River Delta. While there were smaller groups in the area before Egypt was founded, it took the first great pharaoh to unite them all into one civilization. Advancements in agriculture played a big part in the lasting power of Egypt. The civilization did extremely well in cultivating arid land, as well as trading with neighboring city-states to build a healthy economy and prosperity for its people. They were obviously very talented at construction. Egyptians not only built the Great Pyramids, but also many temples and obelisks that marked their territory for hundreds of kilometers.

Thousands of years would pass from one slow stage of technological development to the next, starting in the Bronze Age and culminating into a great civilization, declining through war and other problems, eventually changing into what we see as modern Egypt today.

Norte Chico

25,000 years ago, there was increasing glaciation that yielded an ice age which lowered sea levels. A new land bridge, the Bering Land Bridge, appeared between what we now call Eastern Siberia and North America. The Eastern Siberians achieved the incredible feat of trekking for thousands of kilometers over rugged mountains, in freezing temperatures, facing countless dangers.

This migration occurred over the course of thousands of years. These ancestors from Eastern Siberia colonized North America and eventually became known as the North American Indians.

Some tribes continued traveling south into the South American continent. Around 4000 B.C., many settled in the area along the coast of what is now north-central Peru – this became the Norte Chico civilization, which emerged about a thousand years after the Sumer civilization. Initially, the Norte Chico people had virtually no art or use of symbolism. They were a very peaceful people, as no evidence of the use of weaponry has ever been found. The grandest of the Norte Chico's achievements were their many monuments, such as their platform mounds and step pyramids, made possible by advanced ceramic techniques. There is also evidence that a rather complex government was in place for many hundreds of years to manage the diverse population.

The demise of the Norte Chico came about slowly over hundreds of years through migration of people from outside the civilization. The knowledge and ceramic techniques the Norte Chico developed were taken to other lands by groups that quickly grew into their own nearby settlements. A thousand years would pass before another great civilization was built in the area, the Chavin, and eventually the Olmec, eclipsing the original Norte Chico people.

Olmec

The Olmec came to the area around 1200 B.C., thousands of years later than the Norte Chico, though they were the first to settle in the northern Mexico area. They reigned in the Mesoamerica region for a couple of thousand years before mysteriously dying out. Historians are not certain what caused the decline, but it is highly likely that it was caused by drought, or perhaps an internal unrest of some kind.

Many of the Olmec's great works and achievements would end up being used by hundreds of civilizations that followed, including many techniques for crafting beautiful sculptures. Their culture included complex religious and creative institutions which flourished over a wide area. The advent of the Mesoamerican Long Count calendar, writing symbols, and a variety of sports and games (including a primitive form of football or "soccer") was thanks to the Olmec. Sadly, many of the Olmec's works and cultural artifacts were pilfered or destroyed during the Spanish Conquest in the sixteenth century.

Indus Valley

The Indus Valley civilization emerged at the start of the Bronze Age in the area that is now Pakistan and northwest India. The Indus Valley civilization was thriving at the same time as the Sumer and Ancient Egypt, but it was the largest of the three. Metallurgy was their specialty, including the ability to construct multi-story buildings, plumbing systems, and extensive irrigation works. Many of these techniques would later be used by the Roman Empire, and then they were resurrected by modern civilizations.

A likely cause of so many advances in technology in the Indus Valley was the density of people in the cities. As more people were forced together, society had to come up with new ways to handle the problems that came with a crowded living space. Ways to feed the people required innovative uses of wood and metals to irrigate and farm the land. Nearby city-states would often attack to gain control of these consolidated resources, so advanced weapons and defensive structures had to be developed, as well as laws for internal peace and order. Numerous religions formed that consoled a population that was constantly in a state of uncertain change.

China

In my first book, *China and High Roads Beyond*, I wrote about China's staying power as a civilization. China could be said to be the greatest civilization that has ever existed for multiple reasons, including its sheer longevity and also its technological prowess. Ancient China, like other civilizations, had to defend against invasion. To prevent roaming Mongol hordes from taking over their lands, the Chinese built the Great Wall of China over the course of many centuries.

With such a large empire, it is a testament to the strength of the Chinese that they managed to keep their civilization intact to this day. Only the Mongolian Empire was able to secure a large piece of China's northern territory. Chinese civilization today stretches from the Tibetan plateau to the northeastern Mandarin region, and far southeast to the Cantonese villages.

The Ancient Chinese were the first to invent paper, the compass, printing, and gunpowder. Interestingly, the Chinese originally used gunpowder to create fireworks displays for children. It was not until an accident at a storage facility that gunpowder revealed its true potential, both in its destructive capacity, as well as its excavation potential.

The Ancient Chinese also developed music theory with a unique five-tone scale known as the pentatonic scale, in contrast to today's western diatonic/heptatonic 7-note scale. Much of the music theory of Ancient China is otherwise quite similar to the rest of music theory taught in schools around the world today. Their most unique instruments of the time period were made of bamboo, an excellent resonating material. Flutes and reeds were highly popular instruments. Their two distinct languages, Mandarin and Cantonese, also have a four-tone and five-tone differences, respectively.

BEYOND THE FIRST CIVILIZATIONS

Many other civilizations would sprout up after these first ones appeared. Often times new civilizations would appear after a weak one would split from war and attrition, disease and famine, or changes in climate forced some of its inhabitants to migrate to other areas. The rise and fall of civilizations occurred particularly in Europe, until the Roman Empire around 27 B.C. began to unite all of the surrounding warring tribes and independent villages.

From the Atlantic Ocean to the Caspian Sea, the Roman Empire grew to be the greatest civilization the world would see for hundreds of years after its existence. While all civilizations that came before achieved great things, the Roman Empire reached new heights in construction techniques with expansive cities and road networks, the most complex language to have ever existed, Latin, an alphabet still used today, and a great many systems of law.

If it were not for barbarian hordes attacking, and internal unrest caused by the empire attempting to expand beyond its capabilities, it's quite possible the Roman Empire would have been the first civilization to reach what we would define as a modern society. As with all of the civilizations that came before, though, the Roman Empire eventually began to fracture around 117 A.D. with the death of the emperor Trajan.

The Dark Ages began shortly after the Roman Empire collapsed. Progress in resource discovery and economic growth came to a halt in Europe. Centuries passed with human society using nearly the same materials and resources they had used during the previous ages. In fact, nearly every natural metal that exists today, not counting alloys, was discovered before the Middle Ages even began.[2] The refining process and discovery of new alloys was only picked up again around the start of the Enlightenment period in the 17th century.

RISE OF ISLAM AND A GREAT CHANGE

A wide array of differing beliefs, traditions, and rituals arose within these civilizations, including the three most widespread monotheistic religions: Judaism, Christianity, and Islam. Each has transformed the world in different ways, yet the Golden Age of Islam was unique in that it once provided a foundation for scientific study.

Islam originated back in the early seventh century in the Arabian city of Mecca when the prophet Muhammad claimed to have received a revelation from God (Allah). Muhammad was alone in a cave at the time he had the revelation. He shared the story of what he had seen with others, and the religion of Islam began to take root. The story of someone alone in a cave receiving a vision has been told in various ways in many other religions. Stories are powerful tools for directing a society toward a set of goals, regardless of whether the underlying assertions are true or not. Muhammad single-handedly managed to sow the seeds of a worldwide religion with his tales and conquests.

The following centuries were a time of great progress for the Islamic civilization. One extraordinary achievement was a vast mercantile trade network that stretched all the way to China with their sophisticated three-masted caravel ships. Roads were constructed, along which many villages were built, and different kinds of food were grown, creating a variety of goods and wares for sale.

Astronomy and mathematics became a significant public scientific endeavor. Many of the stars in the sky and the constellations have names that were given to them at this time. Muhammad Algoritmi, whose Latin name means algorithm, was a Persian mathematician who helped spread advanced mathematics, including algebra, to the West. These innovations contributed to the Golden Age of Islam which lasted from around the eighth to the thirteenth centuries A.D.

One of the greatest ideas in the religion was the ijtihad, which means independent reasoning. Individuals were taught to think for themselves and to question the beliefs and ideas of others. Few things were out of the bounds of discussion, with the exception of religious texts and laws. Ijtihad focused on the betterment of the whole civilization through the work of individuals. The basis for today's scientific principles was founded on some of the early versions of ijtihad.

Taqlid, on the other hand, is an idea in Islam that focused on the individual's obligations to the religion – and it demands unquestioning acceptance. Taqlid is about conditioning oneself to Allah, and not asking why. Elements of Taqlid would eventually find their way into the Koran, and they are still practiced by

Muslims today. Questioning the ideology is blasphemous and punishable through a variety of laws. Even verifying facts for scientific purposes is often not allowed. The anti-rationalist school called Ash'ari arose partly because some people were so opposed to scientific reasoning. This anti-science movement grew in popularity and even resulted in the banning of teaching of many subjects in schools. As Taqlid continued to outlaw science and critical thinking, progress gave way to decline.

War, pillaging, suffering, and disease contributed to a steady unraveling of the Golden Age of Islam. Greed within the Islamic empire slowly took its toll; great libraries and institutions lost support or were completely destroyed, and the overall quality of life and opportunities of the people waned. A vicious cycle of corruption quickly set in which exists to this day.

Unfortunately, as time went on, taqlid triumphed over ijtihad. Taqlid appealed to those who had true power, as well as to those who wished to maintain the illusion of power. Ideas can be powerful motivators for good, but they can also be motivators for poor decision making, especially when they are not based on facts and reason. Once the process of forbidding critical thought begins, history shows that it can often take a long time to recover from the strife and retrogression that causes a spiral into an age of darkness.

From the start of the Bronze Age to the end of the Middle Ages there is a span of at least 4,000 years. Extend that back by another 5,000 years to include the Neolithic period and its first civilizations, and roughly 10,000 years of civilization is on record. Before civilizations even appeared, Homo sapiens existed for more than *a hundred thousand* years.

OTHER CIVILIZATIONS

Whether through written language or spoken word, civilizations have passed on their inventions, discoveries and culture to generations that followed. Acquiring such a vast reservoir of information that is continually added to and improved upon allows us to overcome barriers to progress. For example, sending a piece of metal (a rocket) to the moon with humans along for the ride could never have been achieved by a single individual. All of the steps required to get to that point took time to discover, develop, and put into place.

Here is a list of some other prominent civilizations in history worth researching:

- ❖ Ancient Americas
- ❖ Ancient Egypt
- ❖ Ancient Greece
- ❖ Ancient India
- ❖ Aztec Empire
- ❖ Babylonian Empire
- ❖ British Empire
- ❖ Byzantine Empire
- ❖ China and its dynasties
- ❖ Incan Empire
- ❖ Minoan Empire
- ❖ Mongolian Empire
- ❖ Ottoman Empire
- ❖ Persian Empire
- ❖ Roman Empire
- ❖ Russian Empire
- ❖ Scandinavia in Viking times
- ❖ Umayyad Caliphate

DEFINING SCIENCE AND HOW IT IS MISUSED

Science: "The intellectual and practical activity encompassing the systematic study of the structure and behavior of the natural physical world through observation and experiment." – Oxford English Dictionary

The methods and procedures that have characterized science since the 17th century consist of prediction, observation, measurement, experimentation, falsification, and the formulation, testing, checking, and modification of ideas. Of these words, I want to call out *experimentation*. We run experiments to determine whether observations agree with or conflict with our initial predictions. We toss out the idea if it does not conform to the body of evidence. Evidence and collection of facts are not disregarded because of emotional appeal of a contrary idea – it is important to reserve emotional judgement.

We have seen in recent history amidst geopolitics that the word "science" sometimes has a negative connotation.

"I don't believe in science and respect my own views about how the world works," some might say to show that they do not accept scientific truths, often for religious or emotional reasons, or simply because the scientific ideas are beyond their field of conceptualization.

Worse still, they may think facts fall within the realm of opinion: "That's your science, not mine." Glancing for a moment at the definition of science will tell you that this is a nonsensical (and often a purely defensive) statement from those who wish to stifle discussion on matters that make them uncomfortable. This disconnect between the inquiry about our world and the rejection of its actual results is both fascinating and terrifying to come across.

We live in a reality that does not cater to our personal whims and emotional appeals, and yet many go about their lives as if the world was made just for them. Especially when something does not go according to their expectations, they think it is either someone else's fault or perhaps God's mysterious will. Fortunately, we do not live in a world that caters to everyone's personal views, or it would be a chaotic landscape none of us would probably want to inhabit. Science helps us to discover our shared reality, regardless of whether we like what it reveals and how we personally fit into its tapestry.

THE AGE OF ENLIGHTENMENT

The Age of Enlightenment was an era of intellectual advancement that flourished from the 1680s through the 1790s. There was a collective goal of exploration and progress in many disciplines, including philosophy, political thought, social theory, science, art, economics and law. Numerous closely knit nations cooperated freely in the exchange of ideas. Innovative thinkers of the period came from many countries, but the majority hailed from England, Scotland, France, Germany, and America. Traveling from one region to another had become safer and easier, which aided the spread of ideas.

During the Age of Enlightenment many great intellectuals rose to international prominence. Some of the most well-known are John Locke, Sir Isaac Newton, Thomas Paine, Voltaire, Thomas Jefferson, Immanuel Kant, Jean-Jacques Rousseau, Montesquieu, Benjamin Franklin, Adam Smith, and Mozart, among many others. What they had in common was a resolve to improve the human condition by understanding the world from evidence-based viewpoints that relied on logic and reason for discovery. Debates that occurred were very different from those of previous centuries that focused heavily on prejudice and superstition. The Enlightenment marked a turning point in history when humans committed to uncovering, proving and valuing truths, using such means as empirical examination, fervent deliberation, inner reflection, logical argumentation, and the scientific method.

One fundamental tenet of the Enlightenment was that individuals have inalienable rights that they are born with. As outlined by the Founding Fathers of the United States of America, the government's responsibility is to protect those rights; rights could not be considered as granted to individuals by governments, because then governments would have the power to take those rights away. Individuals were encouraged to know freedom, prosperity, comfort and happiness. Individuals had freedom to practice their various religions, without interference from the state. Ending religious fanaticism that led to persecution and torture was an imperative to Locke, Voltaire, Jefferson, and nearly all others. Individuals were similarly allowed to compete in a free, *laissez-faire* market economy that would no longer favor the privileged as a matter of course – on the contrary, equal opportunity and social mobility began to permeate all elements of society. The common man was thus liberated, and the predominance of despotism began to peter out.

Just as important as having great thinkers to come up with ideas was a way to retain that knowledge and encourage its use in the future. Books were published in greater number than ever before, encyclopediae and dictionaries were compiled, and great libraries and universities were built to pass the knowledge and wisdom on to future scholars.

Scientists understood how critical it was to remove biases that would otherwise distort the facts. Superstition was thrown aside, and by replacing it with truth, countless discoveries and inventions followed that brought the whole of society the opportunity to learn and discover. Schools became accessible to more than just the aristocracy. Healthful innovations like new medicines, improved hygiene, and safer food storage further improved the quality of life and increased the average lifespan.

The Enlightenment was the greatest catalyst for progress humanity had ever known, and its effects are still felt today.

THE GREAT THINKERS

While science today is largely a collaborative process found in great institutions like universities, before the Enlightenment got started, science relied on individuals to come up with theories in mathematics, physics, and natural sciences. Some scientists worked with a partner, or in a very small group, but, in general, they worked alone. Scientists would solitarily test out their ideas that later would be built upon by others.

Many great thinkers provided the framework to build civilization as we see it today. Some of the more notable of them are introduced next.

Sir Isaac Newton (Classical Physics)

One of the greatest scientists of all time was Sir Isaac Newton. Born on December 25, 1642, in Lincolnshire, England, Newton lived for eighty-five years. His greatest achievement was the discovery of what we now call Newtonian Physics, or Classical Physics. This type of physics explores the macroscopic world we can see around us and through telescopes. The equations of Newtonian physics are still used today for things like celestial mechanics (how the planets orbit the Sun) and how objects fall here on Earth.

Newton wrote *Philosophiæ Naturalis Principia Mathematica*, a set of books in which Newton used calculus to describe the motion of the planets, as well as many other mathematical formulae. The *Principia* laid the foundation for Newtonian physics. It was written in Latin and published in 1687; it was not published in English until 1729, two years after Newton's death, with the title *Mathematical Principles of Natural Philosophy*.

Newton also built the first reflecting telescope to prove that white light is actually a spectrum of colors. He was able to parse out incoming light into separate colors with a prism for the first time. Newton theorized that you could combine the light back into its natural white form, which he successively did with an opposing prism. The basis of electromagnetism and our modern inventions are due in part to this important discovery.

You might also remember Newton's signature apple-falling incident while he dozed off under an apple tree. The incident occurred in the summer of 1666 when an apple literally fell upon his head, sparking Newton's serious contemplation about what exactly causes things to fall. Unfortunately, gravity won out over the tree itself in 1816 when a storm toppled it over.

Adam Smith (Free Market Economics)

Adam Smith was born on June 16, 1723 in Fife, Scotland. A more complicated European economic environment meant that great thinkers began putting increasing research and thought into the economy. For example, Adam Smith thought that a free market was the best economic system for a civilization, and believed that it could be sustained long term – for hundreds, if not thousands, of years or more. He was highly influential in the development of the free market system, though the idea itself had been conceived much earlier. Smith wrote *The Wealth of Nations* (1776), still widely read and respected by economists today. He also wrote papers and books about morality and justice, most notably in *The Theory of Moral Sentiments* (1759).

One of Smith's central ideas was the "Invisible Hand" – the idea that individuals can inadvertently promote the public interest through working toward their own self-interest; we can rely on an individual's desire for personal economic gain and security to promote the economic gain and security of society as a whole.

Thomas Jefferson (Political Philosophy)

Thomas Jefferson was born on April 13, 1743 in Shadwell, Virginia. He was one of the Founding Fathers of the United States of America. He was the main author of the Declaration of Independence. The third U.S. President, Jefferson served from 1801-1809. In 1803, he organized the Louisiana Purchase of land extending from the Mississippi River to the Rocky Mountains, thus more than doubling the size of the USA. Like Adam Smith, Jefferson was a major supporter of a free market system and a democratic society that had minimal government intervention. He favored strong state and local government and weak federal government, arguing that the people would be served better by officials with strong local interest themselves. Jefferson also designed and constructed by his own hands several buildings, including his mansion home called Monticello. Jefferson was a strong proponent of individual, political and religious freedom.

Charles Darwin (Evolution)

Another great thinker and perhaps the most controversial of them all was Charles Darwin, born February 12, 1809 in Shrewsbury, England. His theory of evolution helped future generations to understand the human species and its origins.

Before Darwin's work, philosophers were largely guessing at how we came about on Earth.

Darwin spent years traveling the world, including five years on the Royal Navy ship the HMS *Beagle*. While the others on the ship charted coastlines, Darwin spent his time on land investigating geology, ecology and zoology. He also collected samples to send back to England. The fossils and other examples of biodiversity, along with the journal he kept of his travels, created much excitement back home.

As the ship sailed from continent to continent, Darwin began to see patterns in and connections between the processes of the planet and the biology covering it. Some of his observations did not make sense to him at first. For example, he found identical species on separate continents and wondered how that could be. He noticed plants and animals that were similar, yet still different enough to call a separate species. His genius helped him notice similarities and differences, as well as how the surrounding environment played a role in shaping them. He observed that tortoises' mouths were shaped just right to consume the available food, and the larger and tougher the nuts they ate, the larger and tougher were their beaks. Darwin termed the biological changes made over time that adapt nature to its environment "natural selection." Darwin then thought about the amount of time it would likely take for biological changes to happen, and the big picture of evolution started to come together.

Darwin also studied steep geological formations like the volcanic rocks on the shore of St. Jago, a large island of Cape Verde off the west coast of Africa. His observations caused him to suspect that Earth was vastly older than anyone previously had thought possible. Our planet had to be incredibly ancient, he figured – a simple concept, yet beyond what most minds would have been able to discern.

Many of Darwin's ideas were written off as preposterous at the time (and still are by some people today). Darwin succeeded by changing his preconceptions to fit the facts, instead of awkwardly trying to fit the facts to popular preconceptions.

Albert Einstein (Early Modern Physics)

Albert Einstein, one of the most renowned scientists in the fields of physics and modern cosmology, was born in Germany in 1879. Einstein showed us how life, stars, planets, and everything else came to be since the Universe formed 13.8 billion years ago. His general theory of relativity outlines how the Universe behaves at large scales, a critical piece of the puzzle of how matter formed after the Universe was born.

Einstein greatly expanded upon the physics laid down by Newton. While Newtonian physics works well enough for describing how the planets orbit the Sun, how gravity works on everyday objects, and what tosses us around when we're on a roller coaster, it doesn't describe very well the properties of the atomic or subatomic realms. In conjunction with Max Planck, Einstein introduced to the world an entirely new branch of physics called quantum physics, which we are still trying to fully understand.

In the subatomic realm, particles behave in ways we cannot precisely measure. Before atoms were seen in an electron microscope, Einstein suggested that particles could spontaneously appear out of nowhere, and that space itself across the Universe is a foam of these virtual particles randomly appearing and then quickly disappearing again. Another oddity of quantum physics includes the pairing of particles across great distances. This coupling, termed "entanglement," would cause one particle to instantly affect the other, regardless of how far apart they were. Einstein famously called this "spooky action at a distance." It turns out that Einstein's idea about entanglement was correct.

Einstein tried to stay in Berlin for as long as possible, mainly to do whatever he could to stop the war effort. In 1933, he narrowly escaped the onslaught of the Nazi war machine and left Europe for the United States. Einstein became an American citizen in 1940.

A few weeks before Hitler invaded Poland, Einstein sent a letter to President Franklin D. Roosevelt to warn him that scientists were starting to think that "extremely powerful bombs of a new type" could be made using the element uranium. The letter also advised Roosevelt that Germany had blocked the sale of uranium from the Czechoslovakian mines it controlled. President Roosevelt took Einstein's words very seriously.

It is hard to say what would have happened to the world if not for this warning. The book and subsequent TV series, The Man in the High Castle, illustrates this alternate reality exquisitely. Consider what would have happened if Einstein, who grew up in Germany, had been indoctrinated with the evil ideology of Nazism? What if he had wanted to work for the German side? Or imagine if he had been unable to flee Germany? They would have forced Einstein to work for them. The irony that this letter to the President of the United States came from a German scientist is weighty. We owe a lot to the fact that Einstein had a good heart. This goes to show the damage that can be done when genius – and the technology it creates – gets coupled with evil.

When asked about his work on the atomic bomb, Einstein often told people, "I do not consider myself the father of the release of atomic energy. My part in it was quite indirect." During World War II, while many scientists were writing about supporting the war effort, no matter what the cost, Einstein wrote about peace. "My pacifism is an instinctive feeling, a feeling that possesses me because the murder of men is abhorrent."

Just before his death in 1955, Einstein signed the Russell-Einstein manifesto, in which eleven renowned scientists implored world leaders to grasp the seriousness of the threat of the proliferation of nuclear arms. Einstein's wisdom shone through when he said, "I know not with what weapons World War III will be fought, but World War IV will be fought with sticks and stones."

Stephen Hawking (Modern Physics)

Stephen Hawking is one of the most famous physicists since Albert Einstein, and arguably as great as Einstein in his contributions to our understanding of the Universe. In contrast with many other scientists, Hawking loves to be in the spotlight.

Hawking got his first chance to peek into the mysteries of the Universe when he began studying physics at the University of Oxford. A few years later, he moved to the University of Cambridge to study astronomy and cosmology.

One of Hawking's greatest contributions to science was his breakthrough idea about the origin of the Universe; he proposed that the Universe started as a singularity (a single point of space and time), working from Roger Penrose's theory about how there's a singularity at the center of black holes.

In 1963, at the age of 21, just as his college years were in full swing, Hawking was diagnosed with ALS, a debilitating motor neuron disease also known as Lou Gehrig's disease. The condition slowly incapacitates the body by shutting down the muscular system, eventually rendering it impossible to speak or move at all.

Imagine being trapped in your mind with the genius of Hawking, but back during the Dark Ages without technology to assist in communicating with the outside world. The disease is a certain death sentence for anyone not living in the age of technology. At the time of his diagnosis, doctors told Hawking he would live for only another three years. Hawking is still alive, 54 years later.

Most of Hawking's work in theoretical physics over the last few decades has concerned black holes. Black holes are objects in space that are so massive that nothing can escape the gravitational pull within them, not even light. Black holes form from the collapse of massive stars at least 5-10 times more massive than our Sun. These stars explode in a brilliant supernova event, blowing off much of their material into deep space. The remaining material quickly collapses in on itself, forming a black hole. Contrary to popular belief, if the Sun were replaced with a black hole of the same mass, Earth would still orbit it in precisely the same way. We would, however, experience a permanently dark sky and much lower temperatures.

In the 1970s, Hawking showed that anything that falls into a black hole is lost forever. However, this theory contradicts everything we currently know about quantum mechanics.

Hawking was the first to recognize this inconsistency; to this day, he works with other scientists to resolve the problem, known as the black hole information paradox.

Hawking also studies the possibilities of alien life and artificial intelligence. In his more recent writings, he suggests that we can expect alien civilizations to be hostile to humans, should we ever encounter them. He thinks it is likely that a visitation from outer space would be in the name of taking our resources, not saying hello. Hawking also suggests that humans are far too slow to adapt to artificial intelligence. He is trying to ring the alarm that A.I. will soon become a threat to humanity.

HOPE AS A CATALYST FOR PROGRESS

Hope can be a powerful driver for change, just as much as fear can be. Hope gives us a reason to invest in long-term endeavors that may not bear fruit until future generations can realize the benefits. As long as we want to build a better future for our children, then we will eventually succeed through hope. Historically, when the future of a civilization was shrouded in darkness, maintaining hope and achieving a sense of purpose as a civilization has been much trickier. The spread of fear throughout the Dark Ages led to despair about the future. Thus little change occurred during this time to progress civilization. During the Renaissance period, the light slowly returned, and hope flourished.[3] Hope allowed enlightened thinkers to feel free to explore new ideas in all areas, and despair about the future was no longer the norm.

CHAPTER 4: THE ENGINE OF MODERN CIVILIZATION

"Modern technology has become a total phenomenon for civilization, the defining force of a new social order in which efficiency is no longer an option but a necessity imposed on all human activity." - Jacques Ellul

For thousands of years, humans were barely able to survive with infrequent sources of food and minimal shelter, and there was almost no progress in technology or technique to secure new food sources and shelter. In many areas of the world during the Bronze Age, the average lifespan was just 26 years. Eventually, humans began exploring their surroundings with greater urgency, and in doing so they acquired new skills that led to the beginnings of civilization.

Many early civilizations adopted methods of agriculture, medicines, and other techniques that set the stage for improving upon these techniques over the thousands of years that followed. We were able to create the industrial world we live in today only after an era of relative stability when new ideas flourished, thanks to an enlightened grounding in science and reason. A single example of the result of this advancement, the average human lifespan, has since tripled.

The Age of Enlightenment brought about refinements in technologies and tools that freed up more time for humans to study what the world could offer humanity. For instance, it no longer took one's entire day to acquire food. Population growth soon exploded, along with an increased human lifespan. With a population that was increasing in number as well as in education level, there were more people with diverse skills who became experts at countless new crafts. This diversification allowed further discoveries in technological niches that eventually led to complex technologies like computers and space rockets.

THE LONG ROAD OF PROGRESS

Progress demands increasing complexity and rates of growth that are difficult to get started, and tricky to maintain. Along with a stable society, there needs to be a means for long-term accumulation of knowledge that can be passed on to the next generation. If knowledge is lost between generations, civilization may never be able to get to the stage of building an airplane that requires the discovery of electricity, plastics, fuel, and so on. Each piece requires a long chain of discoveries and progress that came before.

Another way to think of it is there is no conceivable way a blacksmith in the Bronze Age would wake up one day and be able to build an internal combustion engine – the requisite metals and tools would not yet exist. The blacksmith would be hitting a limit of progress achievable by a single individual. Especially in cases of extreme complexity like the internal combustion engine, hundreds of years of scientific knowledge and preceding technologies would be required before being able to build that engine, even if all the plans were right in front of the blacksmith.

A Scenario in a Galaxy, Far, Far Away...

On planet Noble IV, the people of a budding civilization look up at the stars in wonder of their place in the cosmos. Hanging bright and clear nearly every night are two large moons that shine a pale green. Both are full shaped one evening when a promising young scholar walks out from his forest dwelling on the edge of town. Joshua looks up and sees oceans and inland seas on both moons. Nearby towns had been fighting over resources in an increasingly drought-stricken land in recent years, stealing food, water, and even their women. If only he could build a tall enough ladder to reach the moons, his family might find a safe haven from the chaos around them.

After several months of thinking about the moons, Joshua decided to learn more about a few neighboring tribes that seemed to be friendly towards his own. He had an idea that if all of the friendly tribes could be convinced to work together, they could build a ladder tall enough to reach a moon and escape the local turmoil.

Joshua one morning traveled to a neighboring town with supplies to set up shop. He began making friends sympathetic to his ideas, though he soon discovered that the town already had its own plan for reaching the moons. They worked together over the course of the next several years to gather wood and bindings to build a tower. Eventually the tower reached a height that they calculated was, at most, only a few hundred feet away from the moons. One piece of evidence was that on a clear night the birds seemed to brush the surface of the moon when passing overhead, appearing to take a sip from one of the many lakes before disappearing beyond the tree line.

As Joshua's team tied anchors to a spot on the ground, they quickly discovered that the structural base of the tower was going to require more material to remain stable than they had acquired. Some people suggested using a type of rock in the nearby mountains that could be piled up to anchor the sticks. They spent another year locating the rock, finding just enough that was not too embedded in the mountain, yet they still had trouble reaching higher than the tree tops. The base of the tower quickly grew too wide, needing an exponential amount of more rock for every inch in height that was added.

After a few more years of attempts, most of the workers quit and went off to work on farms, or defend the border lands from increasing hostilities with neighboring tribes. Even those from friendly tribes headed home. Joshua's project that initially looked encouraging ended before it could really begin. He realized that reaching the moons was going to take more than a few friends building a rickety tower. There simply were not enough materials and manpower for the task. He wondered if there ever would be a time when his tribe would reach the moons, and what that era would look like.

Revisiting a Star Trek Movie Scenario

The above story is a good example of people in an era of little technology working together to achieve a far reaching goal (even though they may have failed). In the movie *Star Trek: First Contact*, there is a scenario that provides another example of a group of individuals working to achieve a goal. This time though the technologies exist for the task at hand, but hidden across a world shattered by war.

The scenario plays out just a couple of decades into our future, when humanity is devastated by a nuclear holocaust and has no means of interstellar travel. Even so, a small group of survivors manage to convert an abandoned missile into a spaceship, and one that is capable of faster than light travel. The girlfriend of the designer stresses that it took her months to collect enough titanium just to build the spaceship, and that they would have only one shot at success.

They successfully launch the spaceship years after acquiring the needed materials, and once in orbit, engage the warp drive for a few seconds test. The burst of warp energy occurs at just the right moment to signal a passing alien spaceship. The alien ship diverts its course towards Earth to greet the band of survivors, ushering in a new era for humanity.

Although in the movie they achieve the great feat of building a spaceship, it is still unrealistic to think that one small group could build an interstellar craft without an entire civilization supporting their efforts, especially in such a ragtag way as the movie portrays.

THE GREAT TECHNOLOGICAL LADDER

Until the invention of mechanization, the typical daily life of a family consisted of someone managing the home, while others would work on the farm or go out hunting to collect food. Each generation would repeat necessary tasks day in and day out with little change.

Today, we take for granted the technology that makes our daily lives easier, and perhaps arguably, more enjoyable. We lose sight of the many steps it took to get civilization this far. For instance, you may have gone to the kitchen to get something out of the fridge this morning. For that fridge to exist at all, a series of increasingly complex technologies had to come about first. They start with the metal housing, electrical conductivity to power its functions, and rare gases to drive the coolant. There are also the lights, rubber seals, and a nice colored finish.

You probably switched on a light within the past few hours. A set of technologies had to already be in place in order to get electricity to the bulb. There is the wiring within the building's walls, as well as transmission lines crisscrossing throughout the area, travelling perhaps hundreds of kilometers from a distant power station. That station is processing fuel that was sourced from afar, possibly on another continent. Even the power plant itself is a bundle of technologies with millions of components, requiring years of operator expertise.

Those not versed in construction techniques could still outline the resources and tools it would take to build a treehouse. Even though the details of how to make a comfortable, sturdy one might be unknown, people would have a basic idea of what was required. The details we would not necessarily know could be called "Known Unknowns."[1] We know something is missing, and have a guess as to what it might be. Then there are "Unknown Unknowns," which we know nothing about.

Without the required knowledge of science, Joshua did not have the ability to foresee what it would take to construct a tower of such magnitude. For Joshua, there were too many Unknown Unknowns for him to succeed. Whereas his method of escaping the planet was of course absurd, our civilization does have the required knowledge and capability to go to our moon. To reach this level of knowledge required much time, research, innovation and technological progress that young Joshua did not have. The process is going to be a long and difficult one that only a few civilizations ever achieve, even in the best of conditions. This process of discovery and invention over many centuries in a stable environment will apply to every alien civilization out there as well, including the alien rescuers in the second scenario above.

A technological civilization like ours depends upon many materials and forms of energy to sustain itself and further develop. Let us review some of them so we can better understand why the road of technological advancement is so long and complex, especially if we are to have any hope of maintaining a presence in space long-term.

FIRE – AN ENERGY CATALYST

Every form of energy generation that is usable, whether a biological system, nuclear power plant, or the Sun itself, requires converting materials into other forms in order to extract work energy. This tenet of physics has propelled the evolution of our species, and now the evolution of our civilization. For example, our bodies are able to function by converting raw materials, i.e. food, into an energy source, glucose, a simple carbohydrate or sugar. Once converted, this sugar is then driven to the cells through the bloodstream where it can be directly processed into work energy, including converting food into yet more glucose.

Starting with the simplest of all ways to extract energy out of natural materials, the use of fire through burning wood and other materials was discovered long before humans had any knowledge of other energy types. Once fire was brought to bear, our species learned to use it as a catalyst to extract energy for cooking food, heating the home, and eventually waging war. I say "catalyst" because it is not a source of energy so much as an actual tool for obtaining energy from materials. Without the use of fire, energy in wood, coal, and other non-renewable sources would forever remain locked up in their original and unworkable form.

What if Earth was 100% covered by water, or the atmosphere was filled with gases that do not allow for the combustion of materials? It is conceivable that we never would have even entered the Stone Age if it were not for the controlled use of fire. Fortunately, producing fire is not a problem on Earth, and conceivably not a problem on any planet that has oxygen in its atmosphere and materials to burn.

NON-RENEWABLE ENERGY SOURCES

For hundreds of thousands of years, fire provided a basic means of heating and cooking, until early civilizations found other uses. Wood was the main fuel (and still is in many parts of the world). Wood is easily burned with fire and requires no special preservation. Until the last few centuries, it also grew fast enough to be replenished in just a couple of generations. Our thirst for energy has since far outstripped what chopping down trees can provide.

The Age of Enlightenment brought us to a new energy era in the early 1800s around the time of the Napoleonic Wars. Advanced technologies were being developed across the globe at an accelerating rate, particularly in Europe and North America. Discoveries through scientific investigation became an expected occurrence, and the search for more powerful forms of energy accompanied this growth in technology.

Coal – An Easy Grab

Technological civilization really got its start when humans discovered the potential in burning coal. A lumpy dark substance, coal is the most plentiful non-renewable energy source in the world. In the 1700s, the English found that coal could burn hotter and more cleanly than wood charcoal. Coal-burning stoves and other appliances were soon developed. This marked the first time coal was used for more than generating heat. The very first coal deposits discovered were literally resting on the ground, so it did not take much to gather enough to heat your home for the winter months.

Once the majority of people had a coal stove in the home, easily obtained sources quickly became scarce. Mining techniques were soon developed to get more coal from deep underground. There was also a shortage of steel needed to build ever-expanding urban areas. As industry grew, coal was found to be a great catalyst for the production of steel through the coking process. Coke (not to be confused with the soft drink) is a solid, carbonaceous fuel left over from coal processing. It took a lot of coal to produce a sufficient amount of coke for the smelting process – roughly 770 kg of coal to make just one ton of steel.

Efficiency of the coking process increased significantly over the 20th century through further technological refinements. These changes were expensive to implement, but mining coal sources became increasingly more difficult, so it was an important investment to make. Eventually the process was refined enough to mitigate the initial costs.

Petroleum – The Bubblin' Brew

After we burned all of the easily mined coal, another source of energy was required that was both simpler to store and had a higher density of energy. Petroleum was fortunately discovered right in time for our needs. First there was the easy skimming of oil off of lakes, much like the American Indians did for clothing production. Petroleum also seeped right out of the ground, requiring little effort to capture it. *The Beverly Hillbillies* TV show has an introductory scene where Jed "JD" Clampett shoots the ground with a gun and up comes "the bubblin' brew," as they called it. Yet now, nearly every deposit of oil that can be economically extracted by an individual with readily available tools has been used up.

Around the time that oil was becoming a world energy source, the United States was becoming an economic leader. As the U.S. expanded its reach across the globe, the demand for oil increased. After World War II, Alaska became a leading state of domestic oil production, as well as California and Texas. More recently discovered reserves include the great Sugar Loaf field in Brazil, Azadegan field in Iran, Cantarell Field in Mexico, and several others. Tapping into each new discovery came with increasingly complex extraction methods. Nearly every discovery in the last few decades has been kilometers out to sea and hundreds of feet below the ocean floor. The only new areas yet to be fully explored lie under ice beds in the Arctic.

Natural Gas

While natural gas was a known energy source in China as early as the 4th century B.C., it proved too costly to refine and transport, so it did not get much use until late in the 20th century. The

Chinese were one of the first civilizations to use it to boil water to cook food. They even had a crude system of procuring the gas from seeps in the ground through hollow bamboo reeds.

Today, natural gas is used all around the world, though primarily in the northern hemisphere where it is most abundant and easily transported. Discovering gas reservoirs is similar to finding other petroleum reserves, as they often are situated underground and near each other. Quite often when a rig is set up for oil extraction, they will come up with natural gas first. In the early days of oil discovery, this gas was largely burned off, as it was still too costly to transport. Now we use it to heat our homes and power our electrical generation facilities as a somewhat cleaner alternative to coal and oil.

Tar Sands – The New Oil

Related to conventional petroleum sources, tar sands consist of a combination of clay, sand, water, and bitumen, a heavy black viscous oil. Tar sands are so thick with other materials that they need to be heavily processed. Tons of earth are removed and mixed with water to get to the petroleum-like substance. Even after this process has been completed, the new substance has to be further refined because it is still too thick for anything but literal tar needs, like road pavement. The mined area almost never fully recovers ecologically, even if effort goes into preserving it. Tar sands are still plentiful in northerly countries like Canada.

Factoring in tar sands and other forms of petroleum, current reserves took hundreds of millions of years to accumulate. The process occurred through the natural compression of decayed matter settling mainly on the ocean floor, as well as certain land areas with heavy vegetation. Even some abiogenic sources that do not rely on living matter have been found and used. Regardless of the source, millions of years needed to pass in order to produce what modern civilization will have used in just a few hundred years.

INVENTORY OF NON-RENEWABLE RESOURCES

Humans finally achieved a level of ingenuity to construct buildings reaching higher than the clouds, to fly its inhabitants far over those clouds in metal tubes called airplanes, and to destroy every one of its great achievements in a single nuclear instant, yet we still rely upon energy sources that come from the decayed matter of long-dead organisms. How many plants and animals had to perish to fuel that fifteen gallons of gas in your car? About 1.3 million kilograms of prehistoric, buried organic matter! To create just one gallon of gasoline, about 87,000 kilograms of decayed matter is needed.

How much of all non-renewable resources are left to extract out of the ground, and at economically sensible rates? In searching for an answer to this question, I found many references to *real reserves*, which is quite a different thing from *proven reserves*. Proven reserves are those known to exist, while real reserves are those we suppose exist but have not proven yet. Real reserves are up to four times proven reserve amounts. Below I use the most conservative statistics possible, to ensure that we are well within the margin of error:

Resource	Reserves	Years Left
Aluminum	38 Trillion Tons	60-120
Coal	909 Billion Tons	75-155
Copper	690 Billion Tons	70-125
Gold	165,000 Tons	40-100
Iron	2,563 Trillion Tons	150+
Natural Gas	6,638 Trillion Cubic Feet	35-100
Oil	1.4 Trillion Barrels	45-140

In a matter of just a few centuries, we will have practically used up all of the non-renewable energy and material resources that have thus far fueled the foundation for modern civilization. Some of these resources can be extended by a few hundred years if we are extremely efficient at recycling, and maybe if we get lucky in finding previously unknown reserves. There are multiple ways to estimate where and how much more resources there might be, though, a reason charts like the above exist.

With such quickly dwindling resources, we will face a great challenge later this century to sustain civilization and leave resources for our future children. What is the solution? For energy reserves, there are two options available: renewable sources on Earth and sources in outer space. Eventually it will all need to be acquired from outer space, even if we maximize recycling here on Earth. You can think of our planet as a giant non-renewable ball of surface materials with much of that material forever altered by humans. The same technically goes for everything in space as well, but on such a grand scale that we cannot fathom ever using it all. There are plenty of raw resources in space for trillions of humans.

It has been estimated by NASA and other agencies that the economic worth in today's dollars of all resources within all of the asteroids in the asteroid belt would be equivalent to 100 billion dollars per person. One particular asteroid, 2011 UW158, has an estimated 90 million tons of platinum. Earth's entire known reserves of platinum only amounts to a scant 66,000 tons, or .07 percent of the asteroid's reserves. Do the math on a standard calculator and you'll get an error, as it is too large of a number! Constraining this number to ensure projections are not overly inflated, the estimated value is still in the thousands of trillions of dollars, when directly comparing the value of resources on Earth. This figure does not even factor in all of the other asteroids in the solar system, including the distant Kuiper belt with its many bodies rich with water ice.

RENEWABLE ENERGY SOURCES

Renewable sources are an answer to our future needs for energy, however, the technologies that go with them can be complicated to design and set up, at least for their first implementations. For example, there may be thousands of persons working together to develop technologies that will raise the amount of energy from wind turbines by a few percentage points. Non-renewable energy will have to be used to get the first sets of renewable energy up and running.

We are going to look at five renewable resources, also called "renewables," starting with the more limited renewables, and ending with the ones that are, for all practical purposes, infinite and the most easily achieved with current technology.

Geothermal Power

Geothermal power comes from heat sourced deep underground, usually accessed through fissures and hot springs like the many in Yellowstone National Park. This resource is the most limited of all renewables, with only twenty-five countries having significant sources. Only a fraction of the locations currently have generators installed, as many of the sources are in national parks, preservations, or just not accessible. There are about 73 gigawatts of energy being generated from geothermal power worldwide. In contrast, there are more than 1,600 gigawatts of solar energy being generated as of 2016.

Even though geothermal energy is tricky to harvest, it is theoretically one of the renewables that will survive the longest. It will continue to be a source of power for billions of years. Geothermal energy will even survive the Sun's future red giant phase when the Sun will roast our planet, boiling off the oceans. Earth's internal heat still has to go somewhere, and that will be through its then dry crust in the form of geothermal energy.

Hydropower

The next renewable to consider is hydropower, with an estimated energy potential of about 400-600 gigawatts worldwide. This is more than 50 times geothermal power's capacity. Hydropower uses the age-old method of putting to work flowing water's natural kinetic energy, something civilization has been doing for thousands of years. Even beavers and other animals put up dams in rivers to create barriers that alter the natural flow. Barring an environmental collapse or a sudden ice age, hydropower will last almost as long as geothermal power.

Hydropower is also one of the cleanest of resources, to both extract and to use, so we can be reasonably sure that all sources will eventually be tapped, even those that previously seemed to require too large of a structure to build and maintain. There are only so many rivers and other flowing water bodies in the world after all, and using the ocean currents will be limited to certain coastal areas, at least in the foreseeable future.

The Chinese are known for their massive hydropower projects. The Three Gorges Dam was built along the Yangtze River in the heart of China in 2008, currently the largest dam ever constructed. When the dam was approved in 1992, human rights activists expressed concern about the people living along the river that were forced to relocate, including everyone in two cities with populations over 50,000. In total, about 1.2 million people in more than 116 villages and towns were relocated. Compensation ranged wildly, often depending on the family's connections with dam administrators, but it averaged a measly 50 Yuan ($7) per month for an entire family that may once have owned a significant portion of land along the river.

Biomass

Biomass is fuel that is created from living, or recently living, organic materials. Because these materials grow in nature, they are renewable and sustainable sources of energy.

While biomass has the potential to provide a significant portion of the world's renewable energy supply in the future, currently it is neither easy to produce nor clean. Part of the problem with producing it is that the biomass industry employs a wide variety of technologies, depending on the feedstock, i.e. the source of biomass. Feedstock can take on many forms, including wood, crops, waste, as well as residues and post-processed substances from these resources. Which feedstock is used mainly depends on what is in the immediate area to harvest. After harvesting, there is still a complex series of steps to create the actual material needed to burn.

Another problem with biomass is that it produces a lot of carbon pollutants as byproducts. Of all renewable technologies, nuclear included, biomass is the only one that expels a great deal of carbon dioxide into the atmosphere – roughly 1.5 times that of coal and 3-4 times that of natural gas. The production of biomass also requires a lot of land, which can otherwise be used to grow crops to feed the world.

Because of its many limitations, biomass should be considered a temporary stepping stone on the way to an energy independence built solely on renewables. Think of biomass as a staircase rail. The rail is the support beam to get a step higher, but not completely necessary to get to the top. Just don't hold on to it too tightly, for it may break from supporting your weight.

Wind

Have you ever sailed a boat on a stormy day or tried to swim against a river current? A lot of effort is required to keep stationary, not to mention to make progress. As suggested in the hydropower section, kinetic energy is our friend. Moving water is an excellent source of kinetic energy, but there are limits as to where it can be harvested. This is more or less not the case with wind. Wind can be found everywhere there is an atmosphere, especially in higher elevations.

Let's get right to the downsides of wind, which are almost as few as those of solar energy. Most of the problems relate to how close the wind turbines can be placed to residential areas. Noise can be a problem for some of the older blade technologies, so most wind farms are established kilometers from the nearest urban area. Because of this distance, expensive transmission lines have to be installed and maintained. The distance problem is also caused by some people simply not wanting to see the wind turbines in their backyard. Personally, I find them elegant.

Threats to wildlife can also be a problem with wind farms, though this should be carefully weighed against other considerations. There are birds, including eagles, that get caught up in the blades, and wind farms can threaten the habitats of local flora and fauna. Yet we have to keep the big picture in mind because the benefits of wind power are so great, and the drawbacks are far fewer than media would like to suggest. We do not stop driving cars because the occasional deer gets unfortunately hit (quite possibly by the hunter that would have shot it anyway). In short, wind power's ecological effects are extremely limited.

Once wind power catches on, wind turbines might be a natural piece in the backyard, powering our homes at the cost of a few blades spinning in the afternoon breeze. (There are even some bladeless turbines being developed.) Renewable energy technologies should ideally be self-sustainable to the homeowner, or to the local community, and not reliant upon expensive state-owned transmission lines – wind power could help to provide this independence. Wind power is becoming a relatively cheap and clean solution that has the potential to deliver a significant portion of our energy.

Solar (Photovoltaics)

There is only one other source of renewable energy that truly has the capability to power our entire civilization for an infinite amount of time, and that is the Sun. Thanks to the Sun's abundant supply of energy, there really is no other end-game better than solar power. For those occasions when the Sun doesn't shine, alternative energy sources such as wind power and combined battery technology can fill in.

There is not the need for endless sunny days to produce solar power efficiently. For instance, Germany has a high percentage of cloud cover throughout the year, but it still is a leader in solar power. In fact, Germany produces the most solar power of any country by a significant margin, with a total of 36 gigawatts. Germany is one of the top countries in the adoption of new solar technologies as well.

In addition to coming from a never-ending source, solar power has many advantages. These include a small pollution footprint and low cost, especially with economies of scale.

Yet the true beauty of solar power is its reliability. Once installed, the photovoltaic panels can produce energy for several decades with only minimal maintenance required. The panels can be installed on any building, are not visual marring, completely silent in operation, and they emit zero pollutants.

Cost of the panels are one of the few remaining concerns, and those costs are coming down fast. In fact, many areas of the world are already reaching parity with solar. Parity here means that the cost of converting a particular energy source into usable energy is equal to that of another source. For the first time in solar power's history, it is beginning to outcompete non-renewables. As parity is reached, the adoption of the new source will accelerate. Once everyone has solar panels on their cars and homes, fossil fuels will no longer be needed to engineer these renewable technologies.

Solar, wind, and a sprinkling of the other sources combined would be enough to switch the entire world over to a safe and sustainable energy supply. This would be especially welcomed in Africa where developing countries have intermittent or limited power generation, but plenty of sunlight for solar panels. To ram home the point, transmission issues aside, it would only take 40,000 square kilometers of solar panels with today's technology to power the entire world. This is an area slightly larger than the state of West Virginia. One relatively small piece of land and we could change the world for the better.

INVENTORY OF RENEWABLE RESOURCES

As with the non-renewable sources table, all major renewable sources are listed below. The *amount left* column has been replaced with *realistic potential*. The 1 billion-year mark recognizes the point at which the Sun's steadily increasing luminosity will boil away our oceans. The 4 billion-year mark recognizes the eventual end of our planet when it is engulfed into the Sun during the Sun's red giant expansion phase (or at the very least, burnt to a crisp beyond recognition).

Resources	Potential/year	Est. years left
Geothermal	2,000 TWh	4+ billion years
Hydropower	16,400 TWh	1 billion years
Solar	174 Petawatts/second	4+ billion years
Wave	2,000 TWh	1 billion years
Wind	10,800 TWh	4+ billion years*

*(if the atmosphere survives)

In reviewing the figures above, keep in mind that wave and wind activity is so variable that these estimates will fluctuate depending on journal sources. I also estimated 174 petawatts for solar, which is a truly enormous amount. 174 petawatts is the entire amount of solar energy that hits Earth at any given time. This is far more than human civilization has ever used in total to date, and the Sun hits our comparatively tiny planet with this amount! Put another way, the Sun produces in one second all the energy we would need for the next 500,000 years.

By the end of World War II, just about every earth-based energy source had already been discovered. The next step was to figure out how to extract the bound-up energy. As humans explored more options, new technologies were developed that allowed for more difficult-to-reach raw material deposits to be extracted. New ways of increasing the efficiency of current machines extracting the material were also found. Only fusion and antimatter remain as fundamental sources that are yet to be tapped, because we currently have no easy, safe, and inexpensive way to do so.

THE COMPLEXITY OF TECHNOLOGY

The technology that powers every device today is wildly more complex than anything conceivable just a few hundred years ago. This includes the tiny circuits on the inside of your cell phone to the Gorilla Glass on the outside. And it isn't just the electronics that have seen profound advances in efficiency and durability – our civilization as a whole is incredibly complex, and this is why humanity is special and unique (yet probably still share traits with beings from other worlds).

I'll speed this topic along by focusing on three areas: energy infrastructure and transmission, i.e. "the grid", photovoltaics and energy storage, i.e. "off-grid", and physical goods transportation.

Energy Transmission and Storage

There used to be two competing technologies when electricity was first utilized. They were direct current (DC) and alternating current (AC). While we still use both today in various devices, it was AC that won out for powering our cities. Direct current has a very limited range, so many repeaters would have to be installed into a city's network in order for electricity to reach across town. AC lacks this problem, as the voltage can be ramped up to hundreds of thousands of volts and sent across long-distance wires with virtually no loss in power. This is particularly useful for keeping nuclear power plants away from population centers, and for tapping remote wind farms.

There are more than 764,000 kilometers of power transmission lines that cross the United States. In order to ramp up the AC voltage and then down again once electricity arrives at residential centers, huge transformers have been strategically placed at key points along the lines. There are thousands of these transformers, and most of them cost millions of dollars, take months to construct, and are usually shipped from overseas suppliers. Should a nationwide grid outage occur, and these transformers are permanently damaged, it is estimated that it would take at least several months – possibly even years – to bring the grid back up to 100% capacity.

There are challenges to storing the energy generated by renewables that do not exist with non-renewables. With non-renewables, you have the ability to just keep the original material stored until extraction of the energy is required. This is what we do with gasoline in tanks underneath every gas station. With renewable energy forms such as wind and solar, however, the raw energy must be converted immediately into electricity – there is no intermediate storage medium. If the energy cannot be used at that moment, the electricity generated must then be stored in some other way.

Batteries are the answer. Elon Musk of Tesla Motors aims to create a battery revolution for our renewable energy needs. Musk is one individual that has clearly changed the world for the better with his ideas, including how to power high-quality vehicles with electricity. His team at Tesla Motors completed a battery factory in 2016 they call Gigafactory 1 in the desert sands of Northwest Nevada, where they will be able to build innovative battery packs for both vehicular and home use. Economies of scale will keep the costs down and in turn provide incentive for further growth of the increasingly advanced field of energy storage.

Physical Goods Infrastructure

In recent history, a family could get everything one needed within a couple of kilometers of where they lived. Need food? Just get it on the local farm. Need clothing? Walk down to the tailor shop. This simplistic network of supply diminished as more exotic materials were needed for the mass production of ever-complicated products, and cheaper sources of labor were used to put those products together.

The rare earth metals for your cell phone (which are not really physically rare) are dug up in the heart of China. The rubber for your car's tires probably comes from Malaysia, Brazil, or any number of Southeast Asian countries. In order to get these raw materials assembled into usable products, they need to first be shipped across the world to make it to an assembly factory. Huge container ships travel with millions of products on board, which are then offloaded to container trucks that drive them across countries to their destinations. The system for transporting goods has become extensive. Transportation of goods is one of the largest sectors for jobs today.

Jobs in transportation may become obsolete, however, just as many jobs have become defunct in manufacturing. Automated robots are putting our cars together and smart computers compute new designs for making even smarter computers. Much of American manufacturing has moved overseas where human labor is cheaper. In transportation, it is the drone which will put an end to the need for humans to transport many goods. Amazon.com and other businesses have plans to use drones to deliver packages right to one's door.

Drones help us inspect and oversee areas we would not otherwise be able to reach. Farmers are finding great value in using drones to scan their farmland for crop disease and other issues that could otherwise take days to discover by foot. Drones are also helpful with investigating dangerous locations. Take a drone with you on a hike and I'm sure you'll have fun exploring the terrain! They can even be programmed to find persons who get lost.

While new smart technologies remove the need for humans to be at the front lines, they may lead to a true artificial intelligence (AI) that removes the need for humans altogether. It is reasonable to assume that we will develop thinking machines of some kind in the future and, by extension, machines that do everything for us – both for good and for ill.

EXPONENTIAL GROWTH

Economic growth seems to be a requirement, or at least a strong catalyst, for the ability of a civilization to progress to greater levels of technology. Building rockets and sending astronauts to the moon isn't feasible with a planetary population of just hundreds of persons, or even a few thousand. The efforts of millions are required to support the endeavor and all of its technologies, services, and support structures. Strong growth in our population has occurred mainly because of technological advancement, so the two components go hand in hand in enhancing each other.

Assuming that population growth continues to accelerate, we are going to have to expand into space sooner rather than later, with or without the help of AI. If a town of population 10,000 has a growth rate of just 2% per year, in about 40 years it will have doubled in population, to 20,000. Let's take it to the next level by considering the same population growth for another 1,000 years – suddenly the population is at a billion! No amount of city planning using today's technology would be able to support a billion residents in a single compact region, and this is calculating growth over just 1,000 years, never mind expectations for thousands of years or more.

The entire planet has about 321 million square kilometers of arable land. If all land was used for planting food, we could provide enough food for several trillion persons. That's not really realistic though, so let's scale it down to land in habitable climates, in which case there would be enough land to feed well over ten billion persons.

If current rates of population growth continue to drive the economy, just a few centuries from now, we would need to cover every single land area with solar panels, leaving no room for any other purpose. In essence, current rates of growth will far outstrip available resources, of every kind, leaving only two options available; curtail growth, and by extension our ability to colonize space, or expand vertically in every way we possibly can.

Vertical Farming

If everyone on Earth lived as well as the residents in western society, a couple of Earth-sized landmasses would be needed in order to feed everyone. Fortunately, the most significant use of land, growing food, has a unique new technology that may at least partially solve our problems of land resource overuse: vertical farming.

Vertical farming is as it sounds – producing food on a farm that is vertical, like a skyscraper. Instead of kilometers of farmlands, a city of skyscrapers is built. Each floor of the buildings would be designed to grow specific crops. The efficiency per hectare of food produced could potentially far outweigh what is grown traditionally.

The biggest catch with vertical farming is the difficulty in producing trees that develop deep roots, such as for oranges, cherries, nuts, etc. Currently, vertical farming has found the most success in leafy vegetables, grains, spices, and other small types of produce. Growing larger produce is only a matter of time, though, as awareness of the concept catches on in economic and political circles and the technology further develops.

Here is a graph of economic development and population growth over the last ten thousand years to ram home the point of how we're quickly overcrowding the planet:

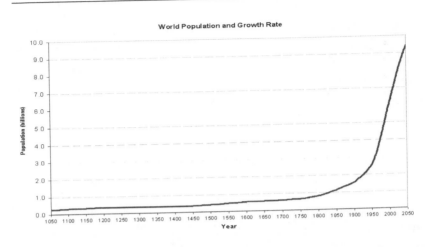

For countless millennia before civilization took root, human population remained less than a million globally, dipping on occasion due to various natural disasters. Today, more are born every year than were born in a hundred years prior to the 20th century. In just the next 25 years, we will have used more energy than in all of past civilizations combined. The rise of our domination of the planet has broken our equilibrium with nature. Even if the population were to remain flat, energy use will still be growing because more of the population is becoming economically wealthy. Millions of cars are being put on the roads and slices of meat on our dinner plates.

BEYOND A SINGLE PLANET

As our civilization started extracting out of the ground easy-to-acquire non-renewable resources in abundance, it increased our dependence on ever more hard-to-reach sources. Whether or not Earth continues to see population growth, as long as its inhabitants continue to use resources in excess of their replenishment, there will be a need to set forth to other worlds. Like the great European explorers that crossed oceans to find other lands, brave explorers of the future will venture into outer space to survive and prosper. If we do not continue to build upon existing infrastructure in order to get into space to discover new sources of energy, the complexity of achieving space travel may lose out to other priorities here on Earth. Once a sufficiently sized colony is established off the planet though, we lessen our dependence upon a single world and its limited resources.

SimCity

I'm a computer gamer, so I would like to provide a game-related example to help paint a picture of how difficult it is to build a minimum level of infrastructure in a new area.

SimCity is a computer game that simulates the building of a city, and it relies on many real-world components that we find throughout our cities. In the game, the goal is to build (and later destroy, if you feel so inclined) a city on an empty plot of land. The land may have mountains, grassy plains, patches of forest, lakes, rivers, and various other types of terrain. You are provided a couple of starting resources. It is up to you to pave roads, as well as designate residential, commercial, and industrial zones for managed growth. Once the population reaches a certain threshold, police, fire, and other services are needed. Power of course is a necessity as well, and it will only become more in demand as the city expands.

In our simulation, we start alongside a river, placing an unpaved road, a couple of industrial zones, and a polluting coal power plant off in the distance. Cars fade in, people move to and from the newly built buildings, and school bells are heard in the residential areas. Fast forward the city's growth by a few decades and you have the entire side of the river covered with infrastructure. There is nowhere else to expand to, except across the river to as yet untouched land. Fortunately, the other side happens to contain even more valuable resources to gather. The city is eager to expand there, but it needs the mayor (you) to build a bridge across the river first.

The river proves a formidable challenge for the city, as the bridge needs to be two kilometers long. Only recently have there been materials available that are strong enough to build such a bridge, and only at great cost. The mayor and council agree to move forward. Funds are reserved over five years with plans and new technologies developed in order to get started. Sacrifices in furthering the quality of life for the residents of the city are made in order to fund the bridge's construction. Construction time is estimated at three years, but various delays, a result of project mismanagement, mean that the bridge is finally completed after ten years. Celebrations kick off throughout the city, and the bridge is immediately crossed by those interested in purchasing cheap plots of land on the other side.

Even though the city has achieved its goal of a bridge to greener pastures, permanent settlement will remain a challenge. After the bridge is completed, funds are further invested to build roads for construction zones. Most of the construction is focused on industrial zones, and the rich mountains in the distance are tapped for rare ores. City funds are nearly drained at this point, though, so building a second police station, school, hospital, and other facilities has to wait. Even with increased taxes that stress both sides of the river, it will be many years before the population on the newly built side reaches the point where it can support itself.

Suddenly a massive earthquake hits the city proper. Not only is power knocked out across the area, but a meltdown is in progress at the nearby nuclear power plant. In a matter of hours, it ends up going critical, spilling radiation across the area, killing a significant percentage of the population. Unfortunately for the settlement across the river, they did not have enough time to build their own power plant beyond basic generators, and health care services are still lacking. Once they realize what happened, anarchy soon follows over fears of how they are going to survive. In a matter of just a few days, the settlement is completely destroyed and the primary city is reduced back to a state not much larger than the settlement at its height.

The bridge symbolizes our venture into space. It took us decades to reach the Moon, explore Mars, and analyze the composition of the other solar system bodies. It will be many centuries more before we can call a travel agency to book a vacation to Pluto, and that journey may still come with dangers. Our SimCity scenario illustrates an obvious need to expand Earth's population beyond the confines of one planet, should Earth experience a catastrophe from which it cannot recover. Simply being in space would allow us to dodge most disaster scenarios fantasized about in Hollywood movies, though it of course would come with some new ones. Most importantly, space has ample room for us to continue expanding.

Related to games and simulations like SimCity, there are many others available. Here are just a few to check out:

- Age of Empires
- Age of Wonders
- Anno Online
- Cities Skylines
- Cities XL
- Endless Space
- Galactic Civilizations
- Masters of Orion
- Rise of Nations
- Sid Meier's: Civilization
- SimCity, SimEarth, and The Sims
- Sins of a Solar Empire
- Spore
- Stellaris
- The Settlers Online

Capturing a Star – Dyson Spheres

If humanity one day manages to colonize a significant portion of the solar system, eventually we are going to need to tap the Sun directly in order to continue to expand. How would you go about directly collecting all, or most of, the energy of a star, might you ask? Getting too close would spell doom for any sort of materials you put in place to capture the energy, so any meddling with the star itself is out of the question. We need something that works at a distance, yet can still capture the star's radiant energy. This is where a Dyson Sphere comes in. Astronomer Freeman Dyson advanced the idea of constructing a sphere around a star to harness its energy. He suggested it would be a necessity for advanced civilizations that have tapped all other resources in their planetary system.

Plenty of variations of the Dyson sphere have appeared in science fiction over the years. One exciting example was in Season 6, Episode 4, of *Star Trek: The Next Generation*. The episode was called "Relics." The sphere's radius was roughly the size of Earth's orbit. The inside of the sphere had an atmosphere just like our planet's atmosphere, and on the surface countless lakes, oceans, and rolling hills. The surface area of the Dyson sphere was approximately *500 million times that of the Earth*; the material required to build the structure would far exceed the raw materials available in the entire solar system by several orders of magnitude. One cannot conceive of how many golf courses could be built upon it though!

In 1964, the Russian astronomer Nikolai Kardashev came up with the Kardashev scale to describe a civilization's stage of technological progress based on the amount of energy it uses.[3] Kardashev's scale had three levels (more levels and extensions were added later by other scientists). Type I uses the available resources on a planet, including the solar energy falling upon it. Type II is capable of harnessing all of its local star's energy (only a civilization of Type II or higher would be able to build a Dyson sphere.) Type III civilizations are capable of harnessing all of the energy in an entire galaxy.

Currently, human civilization would not find a place on the Kardashev scale – we are just not that advanced. Scientists that talk about the Kardashev scale would call us a Type 0; in 1973 Carl Sagan called us a Type 0.7. There do not appear to be too many technological hurdles to eventually achieving Type I. Some scientists believe we will one day become a Type II civilization, but the amount of resources required to harness an entire star's energy may take more raw materials than could ever be obtained.

How likely is it that we would ever reach Type III or beyond? A civilization that is able to convert the energy of a star or a galaxy into a useable form must expel some of that energy as waste heat, and we can easily pick up that signature with current telescopes. So far we have not seen any galaxies giving off the expected signatures that would indicate the presence of a civilization noticeably altering their stars. The Kardashev scale beyond Type II may thus be an unrealistic expectation for civilizations of any kind.

Mathew C. Anderson

CHAPTER 5: A HOUSE OF CARDS
The Collapse of Civilization

"Even with all our technology and the inventions that make modern life so much easier than it once was, it takes just one big natural disaster to wipe all that away and remind us that, here on Earth, we're still at the mercy of nature." - Neil deGrasse Tyson

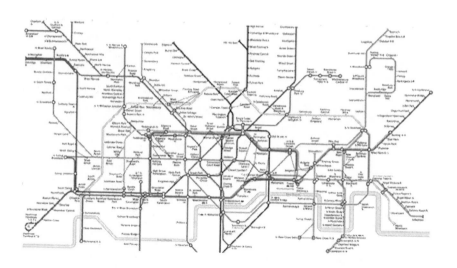

Humanity has traveled a complex and unpredictable road to get to its current technological era. The price that we have paid is our present reliance on technology in an ever more global fashion. The great kings of long gone civilizations have been replaced by multinational corporations that are arguably more powerful than national leaders. Purchasing food from the neighboring farms has largely been replaced by trips to the supermarket to buy food that has been shipped across the country, or even from around the world.

Humanity's achievements have resulted in great feats of engineering like traveling to the moon. Other achievements have also resulted in great destruction, such as the nuclear reactor accident at Chernobyl, which caused tens of thousands of people to die of exposure to radiation. Countless times throughout history, civilizations have faced destruction, but in the last hundred years the pursuit of knowledge and technology has led us to the point that the destruction has the potential to be globally catastrophic. Technology is a double-edged sword for humanity.

In addition to its complexity, today's civilization differs from those in the past in that its energy resources are dwindling. All of the easily accessible resources that a technological society needs are quickly being depleted. Even with technology that helps ease the transition to utilizing diverse, newer sources of energy, many resources will one day be permanently depleted (at least those on Earth). This is a problem that no pre-technological civilization had to the degree we see today. As early as just a few centuries ago, if nearby drinking waters dried up, the population simply moved – a solution not suitable for millions of people today.

Over the course of just a few centuries, humanity managed to invent the technologies that allow us to understand what causes earthquakes, lightning, disease… we have uncovered the nature of the microscopic realm, and even have found ways to explore outer space. As grand as it is to behold, our civilization could be compared to a house of cards that may collapse as a result of any number of significant global disruptions. Think of the top levels of cards as the new upcoming technologies. The bottom row of cards are the raw energy resources we use to power the newer technologies. If the bottom row were to lose enough cards all at once, the entire house would collapse.

Civilization could be destroyed through numerous scenarios, including the mismanagement of natural resources, running out of non-renewable energy sources like coal and oil, natural disasters, and even man-made pollution.[1] In the event of a collapse, every technology that preceded the discovery and development of a future technology would need to be reacquired. If any of those resources are permanently depleted, including manpower, the house of cards we once had may be impossible to rebuild, at least of any degree that it currently exhibits.

CIVILIZATION-ENDING SCENARIOS

To appreciate why humanity should try desperately to not fail to continue its current pace of technological progress, we must consider what cataclysms could affect our way of life. Once those cataclysms are identified, we can then figure out how best to avoid them, if that is possible at all.

Supervolcanoes

Next likely event:	Now to tens of thousands of years
Source of event:	Natural, terrestrial
Potential damage:	Years of cold weather, extinctions
Speed of event:	Hours to a couple of weeks

Let's start with the inevitable – a supervolcano.[2] There are about two dozen supervolcanoes around the world that are, thankfully, dormant. When supervolcanoes erupt, they can be thousands of times more powerful than other volcanoes like Mount St. Helens, Mauna Loa, or even the cataclysmic Krakatoa.

What determines when volcanoes erupt is how long pressure has been building up under the rock that is holding back the gas and magma. The longer the magma is contained and remains in a fluid or viscous form, the greater the potential for an eruption. It's much like when pressure builds up between tectonic plates over time and eventually release that energy as an earthquake. Volcanoes are at least a bit easier to predict, as scientists can estimate how much magma they contain, its viscosity, pressure, and other important signs of an impending eruption.

Calderas are a cauldron-like indentation on a volcano or supervolcano that has emptied its magma chamber in the recent past. There is such a supervolcano caldera at Yellowstone National Park. The last time it erupted, about a half million years ago, it caused over a foot of ash to cover the Great Plains. If it were to erupt today, entire cities would be destroyed for hundreds – maybe thousands – of kilometers in all directions. A single eruption event would effectively end the dominance of the United States in one swift blow, and destroy the global food chain. Our planet would instantly plunge into a mini ice age for decades.

As unbelievable as it sounds, the scenario is real and has happened many times throughout the planet's history. Thankfully, the chances of any supervolcano erupting in the next few thousand years are low, however statistically overdue, as in the case of Yellowstone. There is also the chance that they will never erupt again. On the other hand, a volcano could surprise us all and erupt before you finish reading this book.

While eruptions like Mount St. Helens in 1980 was a major event in our time, it was still tiny compared to even modest supervolcano eruptions. Mount Tambo is a smaller supervolcano that erupted in Indonesia in 1815. High concentrations of ash continued to circulate in the atmosphere for years afterward. In 1816, just after the eruption, global temperatures dropped so low that ice froze on lakes that had never frozen over before. The year was known as the Year Without a Summer.[3]

Asteroid/Comet Impacts

Next likely event: A few thousand years
Source of event: Natural, space
Potential damage: Destruction of the biosphere
Speed of event: Minutes to hours for the immediate impact and fallout event

Asteroids and comets are collections of rock and debris with many that come crashing down to Earth on a regular basis. Once they enter our atmosphere, they are considered meteors. We know about most of the larger objects that could cause Earth serious harm, and none are set to hit the planet anytime soon. The bad news is that there are still tens of thousands of smaller objects that could cause enough damage to set us back technologically for decades. Much like supervolcano eruptions, asteroid and comet collisions are only a matter of time, although generally rare enough to not lose sleep over.

There were two recent events of significance, both in Russia. In 2013, a meteor about 20 meters in diameter and as heavy as the Eiffel Tower came crashing through the atmosphere at about 60,000 kilometers per hour. It streaked across the sky over the city of Chelyabinsk.[4] The meteor, also known as a superbolide, exploded high in the atmosphere with 20-30 times the energy as that of the nuclear bomb dropped on Hiroshima. The shockwave caused widespread damage to buildings, and injured hundreds, primarily from flying glass. There was no warning of the impending disaster, because the meteor was coming from the same direction as the Sun.

In 1908, a couple of thousand kilometers east of Chelyabinsk, an asteroid with a diameter of 50 meters, or about the length of two railroad cars, felled 2,000 square kilometers of trees and instantly killed a thousand reindeer in the deep forests of Siberia. It is known as the Tunguska Event. There was no way that we could have detected the asteroid in advance, for the same reason that we could not detect Chelyabinsk's, let alone try to stop its fiery descent. Fortunately, it hit a completely isolated area.

There are thousands of such asteroids orbiting along different paths in the solar system. Imagine the power of a rock the size of a typical two-story house – it could destroy an entire metropolitan area in an instant. Putting it another way, the Tunguska asteroid contained the energy of 1,000 Hiroshima atomic bombs, or the strength of today's ICBMs of 10-20 megatons of TNT.

World War III

Next likely event:	Uncertain
Source of event:	Man-made, terrestrial
Potential damage:	Collapse of countries, decades of disruption to rebuild
Speed of event:	Hours to years for the most severe effects to occur

War is one of the few scenarios in which we have at least partial control over the outcome, by virtue of how humans would have started the event. Because there is some inherent control, and because it is not likely to kill off all humans, as in the case of a giant asteroid, scientists do not put war at the top of the list of impending disasters. Interestingly, throughout history, large scale wars have often been a consequence of a natural impediment.

As our civilization grows more complex and the chances of human error and technical faults grow, the odds of an accidental world war increase. Case in point: the only reason two nuclear bombs did not go off in the 1961 Goldsboro B-52 bomber crash over North Carolina was because of a faulty arming mechanism.[5] Unfortunately, the possibility of destroying ourselves by mistake will always exist, as long as technology exists and imperfect beings are at the helm.

Despite the hazard that human error poses, global wars are less likely to happen today than in the last century. Going to war over territorial disputes or clashes of culture occurs less often thanks to economic prosperity, instant communications, as well as the necessity to cooperate internationally. In the past, war was sometimes waged simply because a single individual decided he had had enough of his neighbors. Maturity and respect has lessened the need to push the big red button and monger for war.

Should humans manage to establish colonies on the moon and on Mars, our long-term survival as a species will become more likely because we will no longer have to depend on the well-being of our planet to survive. For instance, if a giant asteroid crashes into Earth, some of humanity will survive if we have colonies established elsewhere in space. It is true that we could still lose badly in a "Space War I." Such a scenario could end our trek into space completely. The more areas of the solar system that we can colonize though, the longer humanity will survive even the most dire of disasters.

Pandemic Outbreak

Next likely event:	Uncertain
Source of event:	Natural, human-engineered virus
Potential damage:	Complete collapse of civilization
Speed of event:	Days to years

Moving along to another scenario just as terrifying as world wars, pandemic outbreaks could wipe us out. We are biological creatures at constant risk of attack from within. The human body is an extremely complex machine composed of millions of parts, and the fact is that these parts are vulnerable. Much of our physiology we have yet to completely understand, including how diseases infect the body and spread to other persons.

Epidemics happen all the time somewhere in the world. Humanity recently experienced a terrible outbreak of Ebola in Western Africa. The disease was utterly devastating in its effects. Since the source was unknown for several months, it was much more difficult to properly quarantine the population and warn travelers of what to avoid to prevent further spread of the disease.

The incubation period for the disease is up to 21 days or more, and diagnostic tests require special equipment and highly trained staff in much protective gear. The key to mitigating these natural outbreaks is quickly finding the source, using aggressive sanitation and isolation protocols of the affected population, and developing a cure.

Normally we can anticipate when a viral outbreak may occur and take action before it becomes a pandemic. Preventing an outbreak is more difficult if it is man-made. Human-engineered viruses can be delivered strategically to a population to result in maximum spread. World War III might erupt if a nation or other group was suspected of developing and disseminating a virus on purpose. Fortunately, engineering a deadly virus is incredibly complicated. Nevertheless, countries would be wise to take great care that diseases do not cross their borders.

Regardless of where a disease comes from, pandemics can appear with such speed that they overwhelm any system designed to contain them. A virus can quickly mutate into a deadlier form, or a normally harmless virus could be engineered to mutate into a deadly pathogenic weapon. If these pathogens are stolen by terrorists and dispersed into multiple areas of the population, an outbreak could cripple a nation, or spread throughout the world.

If we zoom into the world of an ant, we can see that its health can get similarly hijacked by an invading disease. Deep in the Brazilian rainforest there is a species of ant that often comes under the control of a fungus, the *Ophiocordyceps camponoti-balzani*. The fungus infects the ant and seizes direct control of its brain. Soon after infection, the fungus directs the ant to kill itself (after the ant has moved the fungus to a new area more suitable for its growth, of course). There is nothing the ant can do but experience confusion and a sense of losing control of its faculties.

Climate Change

Next likely event:	In progress
Source of event:	Natural, space, and man-made
Potential damage:	Mass extinctions (in progress)
Speed of event:	Long-term up to millions of years

Earth experiences changes in climate on a regular basis with the seasonal swings we experience each year, and with less frequent events like the great ice ages. Our planet's slightly elliptical orbit and its tilt toward the Sun both significantly affect the climate; they change how much and where energy hits Earth from the Sun, which results in how warm the planet gets.

There is also an effect called precession that has a long-term influence on the climate. Just like with the spinning of a top on a table, Earth also has a wobble. As a top loses spin, the wobble becomes erratic and eventually the top itself will topple over (pun intended, and also how the name "top" came about). As Earth's wobble becomes more severe, so will the swings in climate per year become more severe.

Scientists have been able to observe what Earth's atmosphere was like up to 800,000 years ago from ice core samples taken in Antarctica. Each year's atmospheric ingredients are imprinted in layers of ice. Ice core samples show that atmospheric carbon dioxide levels have fluctuated between 170 and 300 parts per million (ppmv). Methane, another even more powerful greenhouse gas, was shown to fluctuate as well. Evidence suggests current levels of greenhouse gases in the atmosphere are higher than at any point in the last 800,000 years.

All of these systems that can alter Earth's climate have potentially adverse effects on life, especially if the changes are sudden. Some species do not survive drastic changes in climate, even on a geological time scale. Earth is currently in the beginning phases of another mass extinction event that is thought be occurring far faster than any other extinction event in recent history, thanks to human activities.[6]

Every year there is a certain number of species that go extinct as part of the natural order of life. Scientists track changes all across the world, and with margins of error, they can estimate when a species is near extinction by a lack of presence in its normal habitat. Sometimes creatures thought to be extinct reappear, but typically they do not. In a 2003 paper, Darryl I. MacKenzie, wildlife research consultant, lays out precisely how detection of species and confirmation of extinction rates can be done with great accuracy. Tiger salamanders and northern spotted owls are two examples he cites of species that are on the verge of extinction.

We can learn about historical mass extinction events by studying fossils, ice core sediment samples, oxygen levels in the atmosphere, and other forms of evidence. Since individual methods of studying the past can be imperfect, it has been very helpful that we have had multiple methods. The results from combined studies generally indicate that in the last few centuries extinction rates have been far higher – up to a thousand times higher, in fact – than in the preceding millennia.

There have been many ice ages and other dramatic climate swings throughout history, but life managed to survive and continue to spread across the planet. Some scientists today predict rather doomful scenarios resulting from the changes in Earth's climate over the past couple of centuries. According to the most dire projections, if carbon dioxide levels continue to rise, the planet could warm significantly enough to cause sea levels to rise by several meters by the end of the century.[7] This rise would be enough to permanently flood coastal cities like New York and Shanghai.

Regardless of how the climate changes in humanity's lifetime, Earth will almost certainly recover from the more dire effects in a few million years. Even if 95% of all species were to disappear tomorrow, it wouldn't necessarily spell permanent disaster for life, as it would be able to eventually evolve into a new set of distinct species. New forms of life would take over the niches left by those that came before. Life might go on as if humans never existed.

Overpopulation and Resource Depletion

Next likely event: In progress, and a near certainty
Source of event: Natural, Man-made
Potential damage: Collapse of most of our civilization
Speed of event: Several centuries (since 1700s)

Civilization urgently needs to find a way to live on other planets or bury its efforts and remain confined to Earth forever. The urgency arises from two currently developing problems: overpopulation and resource depletion. Overpopulation means that too many persons are living in an area which cannot support them. A trillion people could live on Earth if the planet had enough resources to sustain them. As more individuals vie for increasingly scarce resources, conflict and control over those resources is bound to occur.

Everything from fossil fuels to rare metals are being depleted at fantastic rates. Many of them will run out within a generation or two because growing populations are using more of the resources. The challenge is to extract sufficient quantities of resources in space, and in tandem make use of more abundant substitutes here on Earth in the meantime. If the technology can be developed fast enough, we have a chance to not only relieve the problem of overpopulation, but also resource depletion.

The current population growth rates thankfully have a silver lining. Long-term population projections suggest that a leveling off of growth will occur in the late 21st century. It is projected that more women will enter the workforce and will consequently have fewer offspring. Birth control will also become more available.[8] Also, as economic status and general health increases for the population, aging parents will not need as many children to take care of them later in life.

Electromagnetic Pulse (EMP)

Next likely event:	Within a hundred years
Source of event:	Natural, space, nuclear weapons
Potential damage:	Complete destruction of electronics
Speed of event:	Instant, with permanent effects

An electromagnetic pulse (EMP) is a short-duration burst of electromagnetic energy. EMPs can be catastrophic when they come into contact with vulnerable electronics.[9] There are two main sources of an EMP that can do great damage – nuclear bombs and the Sun. They differ slightly from each other, but if the pulse is strong enough, the destructive effects (permanent shutdown) on electronic and electrical systems will be similar regardless of the source.

The long electrical transmission lines that crisscross the nation provide an opportunity for an EMP to build a surge that travels along the line to sensitive electronics at the end. When the pulse travels through the atmosphere, then over electrical lines, it creates a voltage surge, which goes into walls of buildings and meets our electronics. No explosion occurs, just an instant silent shutdown of all electrical and electronic devices. (Transformers along the electrical grid, however, would overload and could explode.) Even battery-powered items like flashlights, as well as electronics that are not plugged in, would become inoperative.

One way to prevent the destruction of equipment is to have it "hardened" – a military term meaning protected against attack. Hardening of equipment or structures includes adding metal, concrete, or other materials that redirect the EMP's energy around the object, preventing the energy from entering the internal area. For example, lightning rods harden all major skyscrapers. When a nearby lightning strike occurs, it will be attracted to the rod. The rod is connected to a wire that feeds into the ground at the bottom of the tower. The energy from the lightning strike will be channeled along this wire and dissipate into the ground, instead of into sensitive electronics within the building. The electrical grid can be hardened by using more fuses that stop an EMP from progressing too far along wires.

Another way to protect equipment is to store it in what is called a Faraday cage. Faraday cages are containers designed specifically to protect the items kept within from electrical disturbances. The first Faraday cage was created in 1836 by Michael Faraday, English scientist. The cage contains metal that causes an EMP, or any sort of electromagnetic radiation, to disperse around the cage's exterior.

The world got a direct taste of the power of an EMP when the United States and Russia were testing nuclear bombs in the 1960s. During the Starfish Prime nuclear tests in the Pacific Ocean in 1962, scientists were taken by surprise when electrical equipment was damaged hundreds of kilometers from the blast area. Similar effects were seen in other tests, even with certain non-explosive devices emitting an EMP. The tests clearly showed that an EMP has global destructive potential far above and beyond the immediate effects of the bomb blast itself.

The destruction caused by an EMP would be even more widespread if it came from the Sun. There are major flares being discharged from all across the Sun's surface on a regular basis, but every few hundred years a massive one is directed toward Earth. The most recent flare of significance resulted in the Carrington Event back in 1859. The EMP was strong enough to cause telegraph operators to suffer burn injuries as they handled overloaded wires. The EMP came from an X45-class solar flare, one of the most powerful ever recorded.

If an EMP from the Sun of the same scale as the Carrington Event were to hit today, it is hard to imagine the destruction that would occur. In 1859, very little was powered with electricity. They still used gas lighting – it would take another 20 years before the lightbulb became of practical use. Today, we are utterly dependent on electronics for almost every facet of life, from buying food to filling our gas tanks, from staying in touch with those we love to getting across country to see them. Radios, cars, and everything else we depend on for communication and transportation would be at risk of failure. A solar EMP could cause failure on a global scale that very well could be permanent. There is no other known event that would cause our way of life to collapse so completely, and yet leave most physical items seemingly untouched.

In the next few decades, it is the nuclear bomb that is the most likely source of a EMP strong enough to cause catastrophic damage. Since the Cold War, there have been thousands of nuclear weapons in Russia and the United States collecting dust in bunkers. Many could be modified to enhance their EMP effects upon being detonated. An EMP disaster might occur if a rogue nation or other madmen were able to launch a modified warhead high enough into the atmosphere over a country.

Nuclear Holocaust

Next likely event:	Uncertain
Source of event:	Man-made, terrestrial
Potential damage:	Collapse of civilization, radiation hazards for thousands of years
Speed of event:	Hours to weeks for the most devastating effects to occur

There are few scenarios as horrifying as an all-out nuclear war, if only for the fact that we would have brought utter ruin on ourselves. The projected effects on humanity and almost everything on the planet are so dark and gruesome that they are difficult to comprehend.

Nuclear bombs come in a range of sizes, shapes, and yields (explosive power).[10] They can be fitted onto missiles that can travel to countries on the other side of the world within a matter of minutes, launched undersea from large missile silos on submarines, or high in the sky from bombers launched from fixed land bases. There are enough nuclear weapons – about 14,800 as of October 2015 – between the United States and Russia alone to completely destroy all of the world's cities several times over.

The very first nuclear bomb exploded on July 16, 1945 at the Trinity test site in the Jornada del Muerto desert in New Mexico at a top secret army base. The bomb, designed by a team led by J. Robert Oppenheimer, had a yield of 20 kilotons of TNT, or about the explosive strength to vaporize everything within a few kilometers. The heat blast was felt as far as 160 kilometers away, and Oppenheimer's team reported it as feeling "as hot as an oven." Oppenheimer described the explosion as "terrifying."

Oppenheimer is well known for quoting a Hindu scripture passage: "Now I am become Death, the destroyer of worlds."

Just a few weeks later, two nuclear bombs were dropped on the Japanese cities of Hiroshima and Nagasaki, effectively bringing World War II to an end. The Hiroshima bomb had a blast yield equivalent to about 16 kilotons of TNT, less than the Trinity test itself. It had the codename "Little Boy," it was a gun-type of atomic bomb that used uranium-235, and it killed more than 100,000 persons. The Nagasaki bomb was slightly more powerful, equivalent to about 21 kilotons of TNT. It had the codename "Fat Man" and it was an implosion-type bomb that used plutonium. There were fewer casualties in Nagasaki because much of the city was shielded by mountains, but tens of thousands still lost their lives in the blast.

Russia created the largest nuclear bomb ever, called the Tsar Bomba. It had a blast equivalent of 50 megatons of TNT, which released 1,000 times more energy than the Hiroshima and Nagasaki bombs put together. The designers originally had wanted to make it 100 megatons, but it would have caused significantly more fallout, plus the plane delivering the bomb would not have been able to escape the blast radius in time.

In 1961, the Russians detonated the Tsar Bomba over Novaya Zemlya, an island chain north of the Russian mainland. The mushroom cloud rose for several kilometers above the ground before flattening out. More than 200 nuclear test bombs have been detonated at Novaya Zemlya, producing explosive energy equivalent to 265 megatons of TNT, roughly 130 times the energy of all of the explosives used in World War II combined.

The damage that 50 megatons of TNT could do to a city is astounding; this yield would be enough to completely vaporize everything within a radius of 60 kilometers. It would cause third-degree burns 90 kilometers away. New York City is about 25 kilometers in radius. If the Tsar Bomba were dropped over Manhattan, the blast would obliterate everything in the entire city and for kilometers beyond. The heat from the blast would be felt more than 250 kilometers away. The flash of the blast would cause instant blindness in millions of people. Even if you were hundreds of kilometers away, if you were looking at the blast site at the time of detonation, you could go blind. The observers of the Trinity Test wore special goggles that shielded their eyes from the intense brightness.

While the power of a nuclear weapon to instantly obliterate is chilling to realize, the radiation fallout is, in my opinion, the most damning part. Radiation is completely silent and invisible, and avoiding it is close to impossible (unless you're already in a shelter hardened for radiation when the blast strikes). If a nuclear bomb were to go off near the ground, it would kick up radiation-soaked soil and debris – this material is called fallout. That cloud of material could travel on winds for hundreds – possibly thousands – of kilometers in whatever direction the winds happened to be moving at the time. The fallout is more harmful to life than the radiation released by the bomb into the atmosphere because the fallout remains in the environment for hundreds, if not thousands, of years.

If you were outdoors near the blast, and weren't killed by the blast itself, radiation could cause you to die within hours. Radiation causes damage to DNA and subsequent cellular degradation. Imagine tiny laser pulses poking holes in every cell of your body. Blisters may appear on the skin. The immune system starts to malfunction, so if death hasn't already occurred, disease would quickly set in to finish the job. As time goes on, deformity of the extremities results, as well as blindness, hair loss, and, if you survived for years, severe forms of cancer would become a significant possibility.

If a global nuclear exchange occurred, the entire world would be a minefield of radiation hazards that humanity would have to carefully traverse for thousands of years. Water sources and farmlands would be poisoned by the radiation. Nuclear winter would ensue, with global temperatures dropping for years, killing off crops that managed to survive the initial radiation hazards.

Let us hope that the nuclear weapons used to end World War II are the last we ever do use, and that our civilization never falls from the pure folly of self-destruction.

Other Catastrophic Events

Most of the catastrophes in this chapter are Earth-based. For other space-based scenarios, there are Gamma Ray Bursts and Supernova. Gamma Ray Bursts (GRBs) are powerful, energetic, pointed blasts that come from the death of massive stars. Supernova are similar but they are not pointed – they blast out energy in all directions. Both typically occur in a galaxy only every few centuries or so. Gamma Ray Bursts and Supernova are extremely unlikely to affect Earth in the next few hundred thousand years.

Humans are capable of preventing many of these catastrophes from occurring more than ever before, yet the general population doesn't really rally around trying to do so. Instead of holding a long-term view of existence, we focus our attention on the immediate day's activities with only mild interest in the news of yet another deathly event around the globe. Our nights are restful, and when they are not, it is likely not because the Sun is going to burn up Earth in a billion years, or that there might be a supervolcano going off at some point.

THE FALL OF CIVILIZATION

What would the aftermath of an apocalyptic event look like?

For the first couple of days there would be mass confusion as everyone tried to figure out what had happened. Hope would be high that someone in authority would figure out how to bring things back to normal. Hope, however, would not last long. Soon it would become clear that calamity without precedent had befallen global civilization. People would wonder if they had been attacked by another nation, or by terrorists... fewer people would suspect the natural sources. Depending on the extent of the damage, what caused the event might never be known.

It wouldn't be long before people would loot, starting with stores, and continuing with nearby homes. After a week of small groups of the bravest people attempting to restore order, they would realize that no one was coming to rescue them. Local officials might start by dispersing resources, but then many would eventually abandon their efforts in an attempt to escape the cities with their families. Gangs would form and assume control over neighborhoods. Potable water would be scarce. Fires would burn unchecked. A can of beans would become the new gold. After just a few weeks, the fear of starvation would bring about an ugly new threat - the possibility of cannibalism.

Mass death would begin to occur in phases, starting with the helpless in hospitals, prisons, and other restricted care needs. Disease would set in from lack of medicine and malnutrition. Inability to bury all of the dead would spread more disease. A third of the population could die off in just a few months, and ultimately a majority of the population could expire.

Civilization would essentially be thrown back to the Stone Age within a year.

Putting the scale of disasters into perspective, take the victims of hurricane Katrina. They experienced chaos and fear as soon as the hurricane cut power to the city and the flooding began, but unlike after an EMP, they had radios, phones and other portable devices to help them communicate with emergency services. They also had the world's resources available, yet weeks still went by before they got access to clean water. Even with fully empowered search and rescue teams, hundreds still died.

REBUILDING THE ENGINE OF CIVILIZATION

"We can't just reuse the old designs," John Balboni, an engineer at the NASA Ames Research Center, said in an interview with *The Verge* online magazine on Dec. 5, 2014.

Balboni was looking to recreate and hopefully enhance the Apollo heat shield, a layer of material on the Apollo spacecraft that protected the craft from intense heat during reentry into Earth's atmosphere. He discussed several ideas with other scientists in 2006. The engineers that designed the original heat shield back in the 1960s left detailed information, but most of these engineers were no longer alive in 2006 to explain the details. Balboni's team identified several areas of confusion that led them to wonder if they would ever be able to reverse engineer the old heat shield designs.

"It's like trying to make a cake from your grandmother's recipe. She could leave you all the ingredients and all steps, but you're not going to make a cake as good as your grandmother made," Balboni explained.

Balboni made a powerful statement that says a lot about what it takes to develop complicated technologies. Many of the technologies developed at NASA will likely take many years of work with hundreds of scientists involved, not to mention millions or even billions of dollars of funds. Like the story in the last chapter about Joshua's attempt to reach the moon, we are not going to build spacecraft with one or two individuals loitering around the village wishing well and just thinking really hard about the problem.

The Recovery Stage

Even the worst catastrophes leave survivors to contemplate the future that awaits them. A few years after an EMP event, order would be restored in pockets around the world. The TV show *Revolution* depicts what life would be like in the years after an apocalypse. The stability of the great nations would be no more, and after years of anarchy, unstable despotism would take its place.

Survivors might attempt to rebuild society and reconstruct the lost technologies... it just might take hundreds – or even thousands – of years to do so.

Technologies would first be redeveloped for growing food, building shelters, and setting up a system of trade and government. Once a civilization restores order, progress can accelerate. Design manuals would be found in the rubble of the old cities, original creators might still be alive to advise the new generation, and many machines would be reverse engineered. Perhaps even new and better machines that were not thought of before would be crafted. Eventually the despotism would give way to more civil governments.

Along with loss of life, technology and resources, lack of interest in rebuilding civilization might be a reality for most survivors of an apocalyptic event. Few are going to be concerned with this long-term goal of rebuilding after losing a way of life that has little hope in ever seeing a return in their lifetime. After all, it is easier to destroy than to build. Indeed, many in today's already advanced society have reasons to exist other than a direct intent to build a better future for all of mankind.

OUR UNDETERMINED FUTURE

If we wish to not stumble back into a new dark age, we must do everything humanly possible to avoid the fall of civilization. On a global scale, this means securing our aging power grids from natural disaster or attack, stockpiling key resources such as food, medicine, and water in protected shelters in all major cities, and hardening more military and civilian equipment such as radios, vehicles, and generators. Countries also need to work together to prevent a global catastrophe.

If we fail now to reach outer space and tap the plentiful resources there, humanity may be doomed to a limited existence on Earth, reduced to a patchwork of wandering groups in a hostile and energy-starved environment. Subsequently, we may plateau at a stage of development much more primitive than our current one.

Civilization could be set so far back that we would never be able to travel into outer space again. Children in the new world would read about the old ideals through tattered history books, flipping past images of great cities and rocket ships from what was once a grand civilization largely devoid of the problems they now faced.

Idiocracy (2006) is a satirical science fiction comedy movie set in a dystopian future 500 years from present day. All of the modern conveniences are exaggerated, such as microwave meals, easy access to goods, and a sense of individual freedom and superiority. Society is seen to have long ago reached its peak of intellectual greatness, now instead being pictured as a dumb society reliant on the system for literally every facet of life. Intellectualism is viewed as a threat, and technological progress has screeched to a halt. While life continues on, it is evident that eventually the species itself is destined to having a reduced intelligence, progressing within a handful of generations to a stage not much smarter than the great apes themselves.

Whether it's an immediate disaster, or the slow decline of our intellectual capacity ('intellectual atrophy' as I like to call it), I personally do not want to ever find myself in that kind of crumbling world. What a tragedy it would be if civilization never recovered from a natural disaster or, worse, a disaster of our own making.

One thing is certain if we ever had to rebuild--for better or worse, the way of life for future generations would never be as it was for those who lost so much the first time around.

The Cosmos as Our Savior

Humanity is currently fighting to stay afloat on the great cosmic barge that is the Earth. Achieving a permanent space presence would present a real jackpot: the resources we can harness from Near Earth Objects (NEOs) like asteroids and the moon. The process of colonizing space can start as a business venture of mining asteroids. Even a medium-sized asteroid contains plenty of water that can be broken down into hydrogen for fuel and oxygen for air, not to mention the trillions of dollars' worth of rare metals. Eventually colonies could be established on Earth's moon, and thereafter on the moons of the outer gas giants. All is conceivable with today's great thinkers working on getting us to this point. Expanding our civilization's presence in the solar system – or maybe even throughout the galaxy – will not be easy or quick, just as it wasn't for the first explorers who crossed the Atlantic Ocean. The exciting part is that the possibilities are endless, just extremely challenging to get started.

While sentient creatures colonizing space may be a temporary blip in the evolution of the Universe, humanity has proven that it can be accomplished at least once. We should refuse to stall our trek to the stars when we haven't even left the proverbial driveway. Exploring the rest of our cosmic suburbia may prove to reveal only empty houses, but it's worth every effort because knowledge of the cosmos is empowering to humanity. As far as can be ascertained, we embody the Universe's best and perhaps only opportunity of leaving a legacy worthy of its existence.

Carl Sagan's saying "We are a way for the cosmos to know itself" could be applied locally in the sense of "We are a way for Earth to save itself." We might one day learn how to prevent natural disasters from occurring. We might also one day learn how to push our planet into an orbit further from the Sun in order to escape the Sun's increasing heat. Moving the orbit of the planet may sound extreme, but it should be viewed as a very large engineering challenge to meet and not as an impossibility.

The International Space Station (ISS) is one of many important stepping stones to reaching other locations in the solar system and beyond. All of the other planets and their orbiting moons will provide valuable construction real estate, as well as raw resources, to help encourage humanity to reach even further. These worlds will allow us to run experiments which would be impossible on Earth. Many of those experiments will be critical to ensuring the survival of future generations of space explorers that will not have the luxury of returning to Earth.

For instance, it is extremely costly to build a telescope and launch it into space. Billions of dollars and years of preparation at a minimum is typically necessary, and there is still great risk of it blowing to smithereens upon launch. If permanent colonies were established in space, telescopes could be built directly there, and be far larger and more effective than anything we could ever launch from Earth. Telescopes could be so massive, in fact, that we could easily peer at the atmospheres of millions of other worlds to see if they have life upon them.

At the very least, space will be impossible to overcrowd. There are enough rocky surface areas in the inner solar system alone to theoretically support trillions of humans and animals, not to mention countless locations for space stations.

If the public views space travel as not worth the risks, then we need to do better to educate the public. Human civilization's time to colonize space is now, for we may not get a second chance. As we will explore in following chapters, this might be equally true for every budding civilization in the Universe. One shot to colonize space is perhaps all that anyone ever gets.

Mathew C. Anderson

CHAPTER 6: EXPLORING THE COSMOS

"I think it would be a very rash presumption to think that nowhere else in the cosmos has nature repeated the strange experiment which she has performed on Earth."
– Harlow Shapley

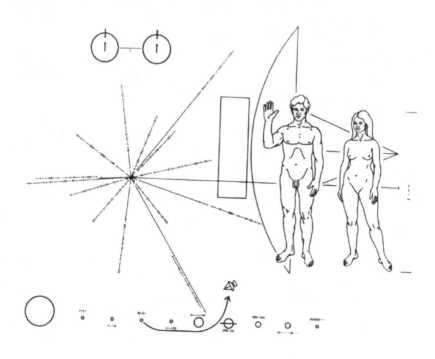

Our latest theories about the origin of the Universe suggest that, in the beginning, all of its energy was crammed into what is known as a singularity.[1] Then suddenly everything, including time, sprang into existence in a "Big Bang." The Universe burst from the womb of nothingness for all of its future astronomers to detect. This initial birth period of the Universe is known as the Planck Epoch, and lasted about 10^{-43} seconds, or a tiny, tiny fraction of a second.[2]

Just as the Universe came into existence, the Planck era ended and another phase of expansion called "cosmic inflation" ballooned the Universe into a much larger size. At about 10^{-36} seconds, the Universe slowed its expansion as it began to cool. While cool by comparison to Planck era temperatures, the Universe was still at millions of degrees Celsius, and an even hotter place than the core of our Sun. At this time the four fundamental forces of nature – the strong and weak nuclear forces, electromagnetism, and gravity – separated out and began to act as they do today.

Microseconds later, as the Universe continued its relentless expansion and cooling phase, protons, electrons, and neutrons formed. The most fundamental building blocks of all matter now existed.

After these first few chaotic microseconds of expansion, the Universe remained in a hot and dense state for the next 300,000 years. Unfortunately, there is no way to know what the Universe would have looked like to the human eye before this time. Everything was still so hot and dense, light itself as we can see it today had not yet come into existence. The Universe would have looked as opaque as a black hole to an outside observer. All of existence previous to about the 300,000-year mark is hidden from our telescopes.

At this young age of 300,000 years, elements heavier than helium could not be created. None of the elements that make up our bodies existed yet; in order for these elements to come into being, the hot and energetic cores of stars would be needed to initiate a fusion reaction that could fuse together heavier elements.

The first generation of stars formed a few million years after the birth of the Universe. They lived short lives of only a few million years, at most. When they died, they exploded, seeding nearby space with heavier elements, and providing the ingredients for the very first planets to form.

THE CITIES OF THE UNIVERSE – GALAXIES

A couple of hundred million years after the first stars formed, an intense bout of new star formation in clusters began to occur, forming the first galaxies. Galaxies can be thought of as the Universe's cities that drive the potential for life and civilization to blossom. Each galaxy contains anywhere from billions to trillions of stars, as well as countless gas and dust clouds. Many of these clouds are still in the process of forming new planetary systems.

While no two galaxies are identical, there are few enough differences to classify them into a handy set of categories.[3]

Spiral Galaxies – Young and Beautiful

Spiral galaxies are host to a large amount of stars, numbering into the hundreds of billions for the largest. They are shaped by gravity into a flat disc of stars circling around a core. This galactic core contains a dense region of stars with the very center harboring a supermassive black hole. Extending out from the center are the arms of the galaxy, giving the galaxy its spiral shape. Currently, 20% of galaxies are spirals. Yet because they are so bright, 70% of the galaxies visible from Earth are spirals.

The spiral shape of these galaxies is a bit deceiving. One might think that it is caused by the rotation of the galaxy itself, but that is not the case. All of the stars in a spiral galaxy revolve around the center in a slightly elliptical pattern. This pattern causes gravitational forces to push together gas and dust in certain areas, igniting vigorous new star formation, which appear as the spiral arms. The largest of the new stars are very bright, so the arms are rather spectacular when viewing a galaxy through the lens of a telescope.

Spiral galaxies also often have groups of stars that are so far away from the core, they tend to follow much more chaotic orbits than the stars in the spiral arms. The farther out from the central pull of gravity, the less pull there is on a star to rotate around the core in an orderly orbit.

The Milky Way galaxy is one of these larger spiral galaxies, but with an added twist: it also has a bar formation of stars that looks like a condensed straight line of these stars running through the galaxy's center. If we could view the galaxy from outside of itself, we would see something like the image above. Not all spiral galaxies have this bar feature, and it was once thought that ours was without it as well.

The Milky Way formed more than 13 billion years ago in the middle of a cluster of galaxies called The Local Group. Two thirds of the way through the galaxy's life, our solar system was born from a collapsing cloud of gas and dust. Ever since then, the Milky Way has carried the solar system in orbit around the center, called the galactic core. The galaxy continues to produce stars; currently the rate of star production is at about seven solar masses per year, which means that seven times the mass of the Sun is produced.

Elliptical Galaxies – The Milky Way's Future

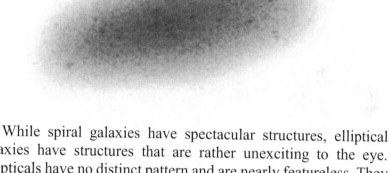

While spiral galaxies have spectacular structures, elliptical galaxies have structures that are rather unexciting to the eye. Ellipticals have no distinct pattern and are nearly featureless. They range in shape from almost spherical to somewhat flat. Their range in size is more diverse than other types of galaxies; from puny little ones with just a few million stars, to truly massive ones with up to several times the star count of the Milky Way, numbering into the tens of trillions.

Elliptical galaxies currently make up about 10-15% of all galaxies. Elliptical galaxies were not a dominant feature of the early Universe – their formation came later. Many of them are the result of dozens of smaller galaxies merging, probably including a few spirals. As the Universe ages, the percentage of elliptical galaxies will steadily increase.

Mergers have occurred several times with the Milky Way already, when dwarf galaxies got trapped in its gravitational pull. Our galaxy has a nemesis on its doorstep, though. Andromeda is one of the few galaxies not moving away from our own. In fact, it's fast approaching on a direct collision course at about 402,000 kilometers per hour. Fortunately, even at this frantic speed, it will still take about four billion years for it to collide with the Milky

Way. After another few billion years, these two spiral galaxies will merge to form one massive elliptical galaxy.

Irregular and S0 Galaxies – The Unwanted Offspring

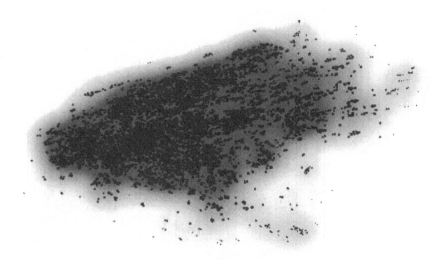

When the Milky Way collides with Andromeda, both our galaxies will undergo extreme changes that will tear stars from their galactic orbits, possibly flinging them far out into space for a time before gravity takes hold and pulls them back. These stars may keep going if gravity loses the tug of war. As you can imagine, changing from two spiral galaxies to one elliptical is going to be a chaotic process. Interestingly, merging galaxies almost never cause individual stars to collide with each other, owing to the vast distances involved. Stargazers on worlds in such an event may never know their home galaxy had merged with another.

Two other less common galaxy types may form as a by-product of the merger: S0 and irregulars. S0 is an intermediate type; its shape is not quite like either an elliptical or spiral, but a contorted version of both of them. S0s contain little gas and dust, have few stars, and are smaller and less bright than most other galaxies. Eventually an S0's shape will become more elliptical or spiral as the stars find a gravitational balance with each other. The merger between Andromeda and our galaxy may result in S0 galaxies forming around the newly merged main galaxy.

Then there are irregular galaxies, which, as the name suggests, have little to no consistent structure. Irregulars have bright regions full of hydrogen that are in the process of forming new stars, sticking out from the rest of the galaxy like a lighthouse on a stormy coastline. Eventually, irregulars will also gain a more defined elliptical or spiral shape.

STARS – LIFE'S PARENTS

Carl Sagan once said, "We are made of star stuff."

This statement is quite literally true. The Universe is composed of roughly 75% hydrogen and 25% helium – everything else is in a sliver of a percent. Stars fuse hundreds of millions of tons of hydrogen per second into helium. As a star ages and uses up its hydrogen fuel, the core of the star contracts in order to maintain the fusion process. The more condensed and now hotter core of the star allows helium to fuse into heavier elements, including carbon and the other elements of which our bodies are composed.

When we detect a star's composition, we are trying to determine its "metallicity" (astronomers call all elements heavier than hydrogen and helium "metals"). When the Milky Way first formed, there were no elements heavier than hydrogen and helium. As supermassive stars exploded, they released heavier elements generated in their cores, seeding the galaxy with them. Future generations of stars contained more and more of these heavier elements. Late generation stars with a high metallicity have a greater chance of hosting planets, as planets themselves require metals in order to form.

Stars of a wide variety of sizes and colors dot the Universe, including extremely massive ones that are larger than our entire solar system. These exceedingly immense stars will only last a few million years, which is far too short a time for life to evolve on an orbiting planet, let alone for life on any planet to build a civilization. Many of the first stars in the Universe were thought to be these massive giants, owing to the higher density of gas clouds during galactic birth.

Our Sun is a typical G-class type star that converts hydrogen into helium, along with a few heavier trace elements. For stars the size of our Sun or smaller, carbon is about the heaviest element that is ever produced.

For larger stars, especially the supermassive ones on the order of eight times or more of the Sun's mass, like Betelgeuse, it's a wildly different story. Elements as heavy as iron can fuse, at least for a brief period of time. Iron literally saps energy from the star's fusion process and never releases it. This prevents the star's energy from pushing outward, allowing the constant inward pull of gravity to overtake. With less energy to continue pushing outward, the star collapses in on itself. As the star begins to shrink, each element separates from the others and settles into a shell shape around the core. First there is hydrogen fusing to form helium, and then lithium, beryllium, boron, carbon, and so on, in layers like that of an onion.

The collapse process lasts just a few minutes before the star becomes unstable and explodes as a supernova in one of the most energetic events since the Big Bang itself. The explosion is so powerful, that in that instant elements fuse to other elements to create heavier ones. Iron fuses into cobalt, nickel, and continues all the way down the periodic table to about element 98, Californium. After that, even the might of a supernova is not enough to create heavier elements – those can only be produced in a lab.

As the supernova explosion progresses, the dying star's outer layers and much of its core are torn apart and sent hurtling into deep space, peppering the galaxy with the gold on your necklace and the copper in your electronics.

The Birth of a Planetary System

In order for comets, asteroids, planets, and every other object in a planetary system to exist, heavier elements than helium are required. Since the early Universe was only composed of hydrogen and helium (and traces of lithium), the very first generation of stars had no rocky planets around them. I'll call these naked stars.

Naked stars are still being born in the Milky Way today, but at a far lower rate than during the birth of the galaxy. This is mainly because most of the gas and dust clouds available to form new stars have already been seeded with heavier elements. As the remnants of stellar explosions like supernovae mix with nearby gas clouds, it will cause the clouds to start collapsing, eventually forming a new planetary system.

These new stars will now contain heavier elements, thanks to the neighboring supernovae. As more supernovae occur, additional heavier elements will pepper nearby gas clouds. At least two other generations of more massive stars had to die before a star like our Sun could be born. It took hundreds of millions of years before metal-rich stars and planetary systems like we see today with our own solar system could exist.

STAR CLASSIFICATIONS – LIFE'S SWEET SPOT

Like galaxies, stars are classified based on their properties, including size, temperature (and thus color), and how they use up their nuclear fuel. About 90% of all stars you see in the night sky are part of what's called the main sequence. Stars in other sequences are either older supermassive stars like red giants, small white dwarfs that have already reached the end of their lives, or neutron stars that are superdense remnants of stars that blew up in a massive supernova.

We want to focus on the stars that are of a stable age with plenty of energy left in them, able to support a planet with life and civilization. The main sequence stars are young adults to older adults – these stars are in the prime of life. Designations of stars in the main sequence range from O, B, A, F, G, K, and M. The Os are the largest and hottest stars, with the comparatively tiny M-dwarfs being the coolest.

Our Sun is squarely within the G-type. It's still a large star compared to tiny M-dwarfs (which are part of the M-type subcategory of stars), but it doesn't come close to the truly gigantic O-type, which can be as wide as our solar system!

O-, B-, and A-Class Stars – Hot and Heavy

The O-, B-, and A-type stars are lumped together here because the chances that any orbiting planets will be habitable for long enough to evolve complex intelligent creatures is close to zero for all three of these types.

O-type is hotter than 30,000 Kelvin, while B-type is between 10,000-30,000K. Type-A broils between 7,500-10,000K. The hotter the star is, the bluer it is. These three classes also have the greatest mass and brightness of all main sequence stars.

A-types appear white to bluish-white to the naked eye and have a mass about 1.5 to 3 times that of the Sun. Examples include the famous Vega star, mentioned in the movie *Contact*.

While all stars have zones where simple life forms could theoretically survive, and probably have the planets on which life can form, A-type stars live less than a billion years before blowing up. A billion years is only enough time for a planet's crust to cool and form a stable layer of liquid water on its surface. Thus A-type stars' lifespans are too short for any planet orbiting them to become habitable. The chances of habitability are even worse around the hotter B and O type stars. Fortunately for the chances of life amongst the totality of stars out there, O-, B- and A-type comprise only about 1% of all stars.

While these giants are not good candidates for the development of complex life around them, they are important for generating supernovae. As explained above, these cataclysmic events are the "heavy lifters" that create and disperse important elements needed for the chemistry of life.

F-Type Stars – On the Edge

F-type stars are classified as yellow-white dwarfs; they are hotter than our Sun, but not as hot as O-, B-, or A-types. F-types are on the edge of being able to support life. Their lifetimes are short – only about 2-4 billion years, depending on their mass. It is not entirely out of the question that life could evolve on a planet orbiting an F-type star, but that life would need to evolve quickly, before the star changes too much to destabilize an orbiting planet's biosphere.

On Earth, it took at least two billion years for life to get to just the multicellular stage. Complex animal life, including the development of the central nervous system, took yet another couple of billion years. It might be that as soon as complex life is forming on planets orbiting F-type stars, the stars are already entering a late stage of development, and heating up so quickly that they end up snuffing out any budding life on their planets.

G- and K-Type Stars – The Sweet Spot

We now get to the most promising life-supporting candidates with G- and K-types. Both are smaller, cooler, relatively dimmer, and longer-lived than the hotter types of stars, giving life a longer period of time to evolve into something interesting. G- and K-type comprise only about 10% of all stars in the Milky Way though. They nonetheless still number in the billions.

While our Sun is commonly known as a yellow dwarf when viewed from Earth, this is a misnomer, as it is actually white when viewed from the undistorted perspective of space. The appearance of it being yellow to our eyes is due to the atmosphere distorting the incoming light. Even within the whole range of G-types, there's only a hint of yellow for slightly less massive stars than our own Sun.

The lifetime of a star similar to the Sun is going to be about ten billion years. Although this is much longer than the relatively brief two billion years of F-type stars, it still doesn't really provide a lot of wiggle room for life to evolve, at least when compared to the coolest of stars. Regardless, ten billion is obviously still enough time for a civilization to form and have a good chance to explore the Universe, as here we are!

The size and lifetime of a star also has a strong correlation to its brightness. Because G- and K-type stars are at the upper size threshold to support complex life, any inhabited planets in orbit will likely have skies not significantly brighter than our own. The average habitable world may actually have considerably dimmer skies, as we take into account K- and M-type stars.

K-type stars are just a bit cooler and dimmer than our own Sun. They have one of the biggest benefits for life to evolve – they last longer than G-types. K-types will last at least ten to twenty billion years. It's plenty of time for life to evolve, be destroyed and evolve again a few times over – definitely enough time to develop a civilization that can reach for the stars.

As a bonus, the habitability zone of K-types would not shift as rapidly as the hotter stars, thanks to how their energy output increases much more slowly as they age much more slowly.

K-type stars are my favorite candidates for life because of the above attributes, not to mention you wouldn't need as strong of a pair of sunglasses. We know that life can exist around G-type stars like our Sun, and K-type are nearly the same. Their planets will have to orbit a little bit closer for warmth, but that won't be a problem for at least the larger of the K-type stars.

M-Type Stars – An Enigma

As we explored in the chapter about evolution, life requires very specific conditions for civilization-building creatures to evolve. This puts into question life around M-type stars.

M-type stars are different from all other star types in just about every way, both with positive and negative consequences for life. M-types mainly include, among others, M-dwarfs, also called Red Dwarfs, which are the smallest of stars in the main sequence. M-types can also be much rarer giants, if they are stars at the end of their life. Focusing on M-dwarfs for the purpose of hosting habitable planets, they have two big advantages: their longevity and their abundance.

The ability of M-dwarf stars to consume their fuel at a slower rate is the main key to their longevity, some of which can last for trillions of years. In other star types, the convection process is limited to either the core or the outer layers, never to the entire star. In an M-dwarf star, the hydrogen mixes throughout its entire structure. As such, the star can last significantly longer before running out of fuel.

In addition to the longevity of M-dwarf stars, the fact that they are by far the most numerous makes them exciting candidates for hosting life on orbiting planets. As many as 80% of stars in the galaxy are M-dwarfs. Even if life were a fraction as abundant around M-dwarfs as other star types, we should still expect to find several times more life-bearing planets around these stars than all other star types combined.

Because of an M-dwarf's tiny size, any planets must orbit very close to keep warm and to retain liquid water on their surface. The unfortunate side effect of the planet being so close, though, is its atmosphere is touched by the star's flare activity.

M-dwarfs are extremely unpredictable in the amount of deadly radiation they produce, at least early on in their lifetime. In a sun-like star, flares develop from convection in the outer layers. With an M-dwarf star, the entire star remains in this twisting and convective dance, so the magnetic field lines become much more contorted, capable of throwing out frequent and extremely powerful flares that can test a planet's magnetic field in protecting life on the surface.

There are other exotic considerations with M-dwarf stars and their planets, which we will get to in the next chapter.

THE STELLAR GRAVEYARD

Stars eventually die. They last from just a few million years, like the O-type giants, up to trillions of years for the flaring M-dwarfs. As they age, the flaring activity calms down, at which point the stars enter a midlife period of relative stability. Our Sun is currently in this midlife stage, at 4.5 billion years of its 10 billion-year lifespan.

The hydrogen the star relies upon for fuel has to run out at some point. Depending on the original mass of the star and how it convects the fuel throughout the star's layers, the fate of stars varies. Their fusion process will eventually stop and for the non-exploding larger stars, what's leftover is a white dwarf star. A white dwarf is a small core that no longer produces fusion, but still shines in the night sky. It will continue to radiate heat for billions of years, to finally rest as a dead black dwarf – a former white dwarf star that no longer emits any significant heat or light.

For our Sun, and stars of a similar type, as they get hotter, its hydrogen fuel depletes, and its core eventually compresses as it starts to fuse helium into carbon. This will heat up the outer layers, causing the star to expand to many times what we see it today. This expansion will come in fits and starts, progressing until its atmosphere cools and turns an orange-red. At its largest, the Sun's radius will be about as wide as Earth's orbit. The Sun will either engulf the planet, or cause it to migrate to a more distant orbit.

The process of expansion will begin long after the Sun boils off Earth's oceans, leaving our planet a dry, lifeless tinder. While complex life will be long extinct, it is possible that the Sun will have shed enough mass during this time period, causing the gravitational pull to weaken, which will in turn allow planets to migrate further out. Our planet may end up nearly where Mars currently orbits. If Earthlings have colonized Mars before Earth becomes uninhabitable, then we might be able to terraform Earth and bring it back to life again.

For the largest of stars, they will form either ultra-dense neutron stars or a black hole. Neutron stars are fantastically dense, and due to conservation of momentum, can rotate hundreds of times a second. The gravity well is so strong that the atoms inside are crushed to the point that the electrons and protons fuse into neutrons, thus the origin of the star's name. A single teaspoon of neutron star material would weigh up to a couple billion tons, or half the weight of Mount Everest!

HABITABLE ZONES

As we know, not all planets are habitable – just look at our own solar system as an example of the limitations around an ordinary and relatively stable star: only 1 out of 8(+) planets can host life. Mercury is simply too close to the Sun, Venus just missed the mark of habitability, Mars had a chance for a while but it is too small to retain an atmosphere, and the rest of the planets are distant giants with a crushing atmospheric pressure hundreds of times that of Earth.

Sometimes called the Goldilocks Zone, the habitable zone is the area around a star, or possibly a gas giant planet, where life has the chance of forming. The boundary begins and ends where liquid water can exist on a surface with atmospheric pressure, i.e. a planet or moon that has a thick enough atmosphere to sustain liquid water. Too close to the star and water would boil off, while too far away and it would freeze.

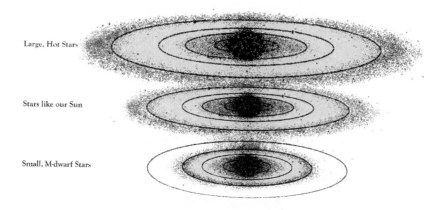

The hotter the star, the wider and further out its habitable zone extends. One of the redeeming traits of F-type stars is their zone is much wider than our Sun's own zone, and yet the star isn't too hot to overly restrict life's chances to appear. The zone of F-type stars is also farther from the stars' dangerous solar flares. A wider zone provides a greater chance for a planet to have liquid water on its surface, and thus be habitable. In a billion years, Earth will no longer have water because of the Sun's increasing luminosity. Every star heats up over its main sequence lifetime, and at ever faster rates for these hotter stars.

For M-dwarfs, the habitable zone is going to be extremely narrow, and very close to the star. Any habitable planets need to practically hug the star, perhaps as close as .10 au, or about 10% the distance from which Earth orbits the Sun. The astronomical unit (au) is the distance from the Sun to Earth – 150 million kilometers, or 93 million miles. We use this measurement to judge distances between bodies in our solar system, including other systems.

Venus and Mars – A Unique Family

Defining the boundaries of a habitable zone has been a heated debate recently, now that we've been discovering so many unique types of planetary systems. A lot has also been learned about how carbon dioxide, methane, and even water vapor act as powerful greenhouse gases that help to regulate a planet's atmospheric temperature. A planet's ability to retain heat with greenhouse gases and a thick atmosphere, will either expand or constrain the habitable zone of other systems.

If Earth and Mars were swapped, it is possible that Earth in Mars' position would still be a wet and warm planet because its larger size allows it to retain a thicker atmosphere, trapping in more heat. For Mars, though, it's uncertain if it would be habitable in any position within the solar system. The smaller the planet, the more quickly it loses its ability to retain heat, generate plate tectonics, and hold on to its atmosphere. If the atmosphere is lost, so will the surface water that is essential for life to evolve. Mars also lacks a magnetic field to counteract the solar wind that would also strip away the atmosphere.[4]

Physically, Venus is comparable to Earth in many ways, yet the planet is devoid of life. When the Sun was 30% cooler at the formation of the solar system, Venus would have been well within the inner edge of the habitable zone, much like where Earth is today within the zone. Instead, Venus is now outside this boundary, too far inward toward the Sun. Our planet has been completely frozen over many times in its past, so a Venus in Mars' orbit might look very much like Earth today, or one of its past snowball periods.

Plate tectonics are critical for a planet to be able to recycle its atmosphere and keep carbon dioxide, methane and other greenhouse gases at balanced levels. Plate tectonics provides a natural thermostat that could allow a planet to survive outside the normal habitable zone boundaries. Scientists think that on planets at the lower end of the mass scale, like Earth, water is a required lubricant to keep the plates moving. Without water, the plates would literally get stuck and the plate tectonic process could shut down. Simulations suggest that a larger planet than Earth may be able to retain active plate tectonics, regardless of its water supply.

The shutdown of plate tectonics is probably what happened to Venus after its surface water evaporated. The poor planet was simply too close to the Sun and any oceans it had boiled away. In addition, the planet's extremely slow rotation (likely due to an ancient collision with a protoplanet), and different internal cooling properties, may have prematurely ended its ability to generate a global magnetic field, which in turn helps in protecting the atmosphere and limits water from escaping into space.

Once Venus lost any surface water it had, the water vapor in the atmosphere started to act as a greenhouse gas. It rose to the stratosphere and split apart into hydrogen and oxygen from the intense ultraviolet sunlight. The hydrogen escaped into space, whereas the oxygen combined with carbon to form carbon dioxide. With plate tectonics non-existent, the carbon dioxide would never be absorbed back into the mantle. The end result would be a buildup of carbon dioxide in Venus's atmosphere to today's level of 96%.

A constant loss of atmosphere is what is happening with Venus and Mars as well. We can even see the atmosphere being blown into space with certain probes we've sent to these planets, such as NASA's VeSpR spacecraft sent to Venus, and NASA's Maven probe sent to Mars. With a non-existent magnetic field that would otherwise keep the solar wind at bay, a strong electric field still exists and may be aiding in the loss of atmosphere and its water. This is especially the case for the now very dry Venus. The electric field pushes up water molecules high into the atmosphere, where they are then carried off by the solar wind.

As a consequence of Venus's failure to retain a life-supporting atmosphere, the planet today is one of the most inhospitable rocky worlds in the solar system. It has an iron-melting surface temperature of 462 C (863 F), winds in the upper atmosphere as high as 400 kilometers per hour, and a crushing atmospheric pressure 92 times that of Earth. The pressure on the surface of Venus is equivalent to the pressure a kilometer below Earth's oceans. Life could very well have started on Venus billions of years ago, but, sadly, complex intelligent life would never have had enough time to develop.

A Galactic Habitable Zone

While there are zones around the Milky Way that tend to produce stars with greater metallicity, and thus also planets, it's not as easy to define as recognizing a planet is within its star's habitable zone. Still, we can calculate some estimates based on a galaxy's structure. Generally, the farther a star is from the center of the galaxy, the less likely it is to have been seeded with heavier elements. Far out stars in the galactic halo are called Population II type stars. Our Sun is a Population I type star and is located well within the galaxy's main bulk of stars.

The habitability of a planetary system within a galaxy is determined by two primary factors though. The first is whether or not a star has enough heavier elements so that planets can form. The second factor is whether or not a star is located in a region of the galaxy devoid of destructive phenomena, such as supernova and gamma ray bursts (GRBs). If both of these criteria are met, then we could say that a planetary system is within the Galactic Habitable Zone (GHZ), at least for as long as it stays in the zone. All stars in the galaxy are moving, so this safe status for any system is likely temporary, including for our own solar system.

Some star systems have more dangerous paths through the galaxy than others. Occasionally two stars will drift close to each other, perturbing their surroundings and threatening any habitable worlds with incoming asteroids and comets. Our own solar system has a cloud of primordial comets and asteroids far beyond the orbit of Pluto, called the Oort Cloud. The material orbits over the course of thousands of years, rarely disturbing the inner planets. When it gets disturbed, though, it can be devastating. At least one of Earth's great extinction events is thought to be caused by a star passing by the Oort Cloud and knocking comets and smaller asteroids into the inner solar system.

Another extinction event in Earth's history is thought to have been caused by the solar system being too close to a supernova or a GRB. Fortunately, supernovae only occur in a galaxy about three times every hundred years, and GRBs occur far less frequently than supernovae – only about three times every million years. The energy from a GRB also needs to be directed towards a planet, as they do not spread out their radiation in all directions as much as supernovae do. If a GRB ever does hit Earth, it would take as little as a few seconds for it to destroy our protective ozone layer, without which nearly all higher forms of life would die. Millions of years would need to pass for life to recover and become as grand as we see it today.

Fortunately for us, our solar system resides comfortably between two galactic arms as it revolves around the galaxy. Like the galactic core, these arms are areas that form more massive stars which will eventually explode violently. Our solar system only rarely crosses these high risk areas. Other systems follow much more chaotic or elongated paths. Some of them will remain in a galactic arm for millions of years, causing habitable worlds to be hit with multiple radiation events, perhaps resulting in their becoming sterile permanently.

Sadly, the neat little arrangement our solar system has with the Milky Way is only temporary. Once Andromeda collides with our galaxy in about four billion years, every star system will be flung in different directions. Some will even be ejected out of the newly formed elliptical galaxy altogether. A galactic habitable zone may become a meaningless term in this chaotic scenario.

ALIEN PLANETARY SYSTEMS

In the search for distant astronomical objects like exoplanets (planets outside our solar system, also known as extrasolar planets), telescopes are a critical tool. The first generation of telescopes used for planet hunting were built in places like Hawaii, on top of mountains, such as the Caltech Submillimeter Observatory (CSO), built in 1985. Since a telescope has to peer through Earth's atmosphere, it needs to be as clear of as many observational distortions as possible, such as pollution, including light pollution. Humidity can also interfere with clarity, which is one of the key reasons that mountain tops make excellent sites for telescopes.

Discovery of the first exoplanet, 51 Pegasi b, was made in 1994 by Dr. Alexander Wolszczan, a Polish astronomer at Pennsylvania State University. He discovered what is called a "hot Jupiter" orbiting 51 Pegasi, a star in the Pegasus constellation. Hot Jupiters are massive planets the size of Jupiter that orbit extremely close to their host star. While the discovery of 51 Pegasi b was exciting news, the planet turned out to not be a candidate for supporting life. Not only was the planet extremely close to its host star, but the star itself was a pulsar. Pulsars are dead stars that previously exploded in a supernova; what remains is a super-dense core that spins rapidly, giving off radio pulses as it rotates. Unfortunately, there is little hope of life around such a hostile environment.

Even though life may not exist on 51 Pegasi b, the discovery of the hot Jupiter proved that other star systems contain planets.

The discovery also proved that at least some of those planets can survive the death throes of their stars, or possibly even be created among the debris leftover from the supernova aftermath.

Just a year later, in 1995, the Swiss team of Michel Mayor and Didier Queloz discovered another planet, but this time around an ordinary star like our Sun. The planet was about the size of Jupiter, and orbits its star so closely that it makes one revolution in just over four Earth days! Its atmospheric temperature is hot enough to melt lead, making it impossible for any spacecraft to survive on the surface for more than a few seconds. Still, it was progress in finding a planet like our own.

The Golden Age of Planet Hunting

Planet hunting is multitudes more challenging than trying to find an actual needle in a haystack. Our ability to confirm planets varies with the size of the planet and how far out it orbits from its star. The smaller the planet, the harder it will be to detect, and the farther out the planet orbits, the less its effect on the parent star can be confirmed.

As planet confirmation took off late last century, scientists, eager to detect the first exoplanet, worried that we may never find a planetary system like our own, regardless of the detection method. Before the first exoplanets were finally discovered, there was an expectation of how a system ought to be configured: the small rocky worlds would reside closer to the parent star, gas giants and icy worlds would inhabit the outer regions, and somewhere in the middle would lie an asteroid belt – a nice arrangement that provided lots of room for a planet (or many) within the habitable zone.

Except that is not what has been discovered.

Discoveries so far suggest just about every configuration possible *but* one like our solar system. The very first planet discovered, 51 Pegasi b, orbits in just a few days around a dead star. Most systems also host super-Earths, which are planets that are just a bit larger than Earth. We have also discovered planets with highly elliptical orbits that make possible any other planets in the system doubtful; as a planet's orbit becomes less circular and more elliptical, the planet has a greater chance of crossing paths with another planet. Even binary stars (two stars closely orbiting each other) have been shown to host planets.

Detection Techniques

One of the first planetary detection techniques used, the radial velocity method, detects the tug of a planet on its parent star. The radial velocity method requires the planet to be extremely close to its star, because we're relying on very tiny readings of how the star is being affected by the planet gravitationally, not through observing the planet itself. This method is very accurate in estimating a planet's size and distance from its star, but it is not able to tell us much about a planet's composition, including what gases make up its atmosphere.

The most successful method to date has been the transit photometry method. As a planet passes in front of its star, the overall brightness of the star dims by a tiny fraction of a percent. The larger the planet that passes in front of its parent star, and the dimmer the star, the easier it is to detect a dip in overall brightness. The Kepler Space Telescope was built specifically for this technique. With it, we have been able to identify thousands of planets with thousands more yet to be confirmed. Confirmation takes several revolutions of the planet to ensure no false positives.

The transit method has an advantage over the radial velocity method in that it can detect planets further from the parent star, out where the habitable zone lies. Other advantages of the transit method are that planets as small as Mars can be detected. The transit method is our current best method for discovering a habitable world similar to Earth.

There are a few disadvantages with the technique. At most, only 10% of all stars and their planets will be aligned in such a way that, from our vantage point, we can see any planets pass in front of the star. Kepler has a field of view of more than 145,000 stars, yet it has only confirmed about 2,300 exoplanets so far.

Another disadvantage is that confirming the existence of planets requires at least three detected orbits. If astronomers are trying to confirm a planet around a star like our Sun at the distance Earth orbits, then it will take a year before a planet will pass in front of the star from our point of view. To detect a planet as distant as Saturn in this way, it would take 29 years! Confirming a Saturn world would thus take nearly a century in Earth years.

Other more advanced techniques include direct imaging of a planet by blocking out the light of the parent star. This approach is one of the most promising techniques because of its ability to image planets directly, regardless of how a planetary system is aligned with our point of view. The technique uses a star shade (thin film of material floating in space) to block out the light of a star, allowing us to see any orbiting planets with a camera positioned behind the star shade.

Another method uses a quirk of physics called gravitational microlensing to peer through a planetary system to image a planet directly. Gravitational microlensing works much like the lens of eyeglasses, except it's a star that's bending the light. The light from a star travels towards another large object, say another star, which causes the light to bend around the object in such a way that focuses that light more than it would be traveling in a straight line. That focusing of the light allows us to more closely evaluate the star's properties, and because planets are close in orbit, find the light reflected off the planet as well.

Whichever method is used, planet detection is a very difficult and sensitive process. It's truly amazing what planet hunters can tease out of the data despite the accompanying interference. For example, to find a typical Earth-sized planet around a sun-like star using any of these methods would be akin to having a firefly in front of a spotlight in San Francisco, while you are in New York and using your unaided eyes to try and see the firefly.

A New Generation of Telescopes

Most telescopes are capable of numerous types of observations, including planetary research, but also Earth based science like meteorology.[5] Scientists bid for their use, especially on the larger and more powerful telescopes. Up until 2009, we had no telescopes exclusive to planet hunting, and certainly none launched into space. Investors didn't want to incur the expense of launching a telescope just on the seemingly slim chance that a few exoplanets would be found. There were means of detecting planets with existing telescopes, so we first used those to make initial discoveries that then justified further equipment.

Once it was clear that exoplanets existed in abundance, the Kepler Space Telescope was launched to discover more. The great news out of the data collected so far is that just about every star seems to host at least one planet, and probably a handful more. This includes binary star systems, as well as smaller M-dwarfs. As detection techniques are refined, we're discovering ever smaller planets than even Mars and, most importantly, at orbital distances where liquid water may flow on the surface.

There are some very exciting telescopes being built, due to start operations this decade, many of which are dedicated to planet hunting. The most sensitive new telescopes are those that will be launched into space. Two upcoming ones will replace existing telescopes with far more powerful instruments: the Transiting Exoplanet Survey (TESS), and the James Webb Space Telescope (JWST). There are also quite a few being built on the ground, such as the W. M. Keck in Hawaii and the Atacama Large Millimeter/submillimeter Array in Chile.

TESS is due to launch in 2017 and is to be the successor of Kepler. It will also use the transit photometry method of detection. As mentioned earlier, this method probes stars for planets that pass in front of their light. The benefit of this method is that we can understand in detail the planet's overall size, mass, water content, atmospheric density, composition, and even any industrial pollutants in the atmosphere.

The JWST is scheduled to launch in 2018, the successor to the Hubble Space Telescope. Much like Hubble, JWST will be tasked with other research priorities, not just planet hunting. The telescope will be the most advanced we've ever sent into space, and also the costliest, at more than 8 billion USD. At several times Hubble's size, it will have to be folded up and then unfurled in space. As you can imagine, the process of unfurling is risky, so ongoing testing is being done in order to get it right. If the telescope malfunctions once in space, it will be so far from Earth, at an incredible 1.5 million kilometers, that it will be practically unrepairable without costly missions that take years to complete.

The JWST will be observing objects in the infrared wavelength of the electromagnetic spectrum. It will therefore need to be far away from our Sun, otherwise it will pick up the infrared portion of the Sun's energy, drowning out any signal of a distant planet. The telescope itself will also have to be extremely cold, otherwise its own electronic heat will distort the image; in fact, it will be operating at −225 °C, or 373 °F below zero!

The best location for the JWST is what's called a Lagrange point. These are points in space between two large bodies, like the Sun and Earth, where the gravitational pull between the bodies is balanced. Place an object there and it will stay there, instead of being pulled more strongly one way or another. There are five such gravitationally stable points around any two large bodies. Only one Lagrange point, L2, will be a suitable location for JWST. For Earth and JWST, L2 is located on the far side of the planet from the Sun, and beyond the moon's orbit.

WHAT WILL WE FIND?

Our sciences have come a long way since Galileo discovered in 1610 the first moons orbiting another planet, around Jupiter. We understand a lot about the basic makeup of the solar system, the Milky Way galaxy, and the farthest reaches of the known Universe. There seems to be a lot of commonality when we start categorizing things like galaxies and stars, but once exoplanets were discovered, astronomers quickly realized how much variety the Universe has yet to reveal to us. We will explore the possibilities in the next chapter.

CHAPTER 7: THE BOUNDARIES OF HABITABILITY

"Some may argue that a diamond is still a diamond, even if it is one amongst millions. It still shines as brightly."
– Guinan, Star Trek: The Next Generation

Stars, planets, and life itself arise from a Universe governed by universal laws of nature. Life on Earth has evolved in accordance with these laws, and alien life elsewhere will do so as well. That alien life may have a stockier body due to its planet's stronger gravity, or larger, yellow-tinted eyes due to a slightly different atmosphere, but there's a reasonable chance it will have a body and eyes. Even different civilizations' inventions will have common properties, owing to the laws of physics, economic constraints, and other universal conditions.

In this chapter we will explore a variety of worlds that may have a chance at being habitable. A continued reference point is our own solar system we've come to know and love. Starting with the inner rocky planets, including Earth, they share many properties. They all orbit in a nearly circular path around the Sun, each has an atmosphere, and a solid surface that one could set foot upon, a day and night cycle, and seasonal weather patterns. They are also all rich in organic compounds, including the chemicals that emerge from active (or previously active) volcanic systems. Life is not found on any planet except Earth, though. These sterile worlds seem to be the norm rather than the exception, even with so many shared properties.

As we talked about in the previous chapter, there are two primary considerations for planets to be potentially habitable (moons will be investigated later): a planet's distance from the parent star and the planet's size. These parameters are important in every system, even binary or trinary star systems. The orbital distance of a planet has to allow for liquid water on the planet's surface. Too close to the star and water will evaporate into space. Too distant – beyond what's called the Snow Line – and water will freeze. Regarding size, most smaller planets within a star's habitable zone, like Mars, will not be able to support a thick enough atmosphere or protective magnetic field, and gas giants have atmospheres that lack many of the minerals thought needed for life to evolve.

Encouragingly, of all the stars we've looked at so far that have planets in orbit, it seems the average number of earth-sized planets found in the habitable zone is at least one per star. The excitement can be tempered with the fact that Mars, and nearly Venus, also reside within the Sun's habitable zone, and one wouldn't want to book a vacation to either of them anytime soon.

Without closer inspection of planets, what we think we know about them can be invalidated with the next discovery. For example, the planetoid Pluto was thought to be a completely sterile world devoid of even the thinnest of an atmosphere, with no geological activity of any kind – even less life-friendly than our own moon. What we in fact found though was that Pluto has an active geological system with ice mountains thousands of feet high, cliffs and troughs stretching hundreds of kilometers, and a thin atmosphere that snows nitrogen in regular seasonal patterns. Even with these active systems, Pluto unfortunately still seems to be a sterile world.

The more we learn about the worlds in our solar system, the more we're both surprised at their variety, and also disappointed at their revealing harsh constraints on where life can appear. This surprise and disappointment duality is likely to repeat itself as we continue to explore other planetary systems.

INTRODUCING SUPER-EARTHS

SUPER-EARTH

EARTH

In between gas giants like Jupiter and the tiny dwarfs smaller than Mars, there is a category of planets larger than Earth, called super-Earths. Super-Earths include both rocky worlds, as well as mini gas giants that are similar to Neptune and Uranus, but smaller.

So far, the possibility of life on these earthly giants is looking both promising and a bit uncertain. The larger super-Earths are believed to have extremely thick atmospheres of hydrogen and helium, near the density of Venus's atmosphere. The smaller super-Earths, however, up to about two times the mass of Earth, are believed to be good candidates for life, despite still having somewhat of a thick atmosphere. The smaller super-Earths are much more likely to have a solid surface as well.

Scientists are close to being able to detect the composition of the atmospheres of these worlds, and thereby gain great insight into their habitability.

Gravity of super-Earths

Once a planet's mass and volume is calculated, we can easily estimate its gravity.[1] Let's assume we discover a planet tomorrow that has the same density as Earth, but with a radius 25% greater. This would put it well within the super-Earth category. It would have about three times Earth's gravity. That sounds crushingly uncomfortable, and it would be to us, but not necessarily for any life that evolves to adapt to that pressure. Fighter pilots can handle forces greater than 3G (three times Earth's gravitational pull). Human physiology only starts to have problems upon reaching 5Gs for more than a few seconds. At about 8-10G, we risk passing out; the heart becomes unable to pump blood to the brain.

There are surprisingly few obvious physical limitations that would prevent a civilization from developing on a stronger gravity world. The most significant issue would be the civilization's ability to send objects into space. Even a modest increase in gravity would noticeably increase the fuel required to launch a rocket into orbit. On Earth, it already costs thousands of dollars to send even a few pounds into orbit, and over 90% of the mass of a rocket is in the propellant.

Other than the interest to explore outer space, a civilization would only need to compensate for a higher gravity environment, such as using thicker steel for skyscrapers, and lighter composite materials for airplanes would probably be researched sooner than they were on Earth.

Geology of super-Earths

The geology of a planet is critical to maintaining a stable environment in which life can thrive. Super-Earths may be even more geologically active than Earth, with vigorous plate tectonics to recycle their atmospheres and keep them cool enough for life to thrive. One might think that a vigorous shifting of the plates would make these planets extremely unstable. The case may be instead that because they can vent their interior heat more frequently, globally impactful events like volcanoes and earthquakes would occur less frequently.

Atmospheres of super-Earths

How exoplanets' atmospheres would (or do) affect any life on them is perhaps one of the most complicated and least understood aspects, even more so than what lies beneath their surface. Until super-Earths' atmospheres can be studied up-close and in detail, models are all we have to reveal the likely limits on life, and those models are currently showing a wide range of possibilities.

Most models of super-Earth atmospheres suggest that the smaller super-Earths have a thin nitrogen-covered envelope, while the larger ones are more likely to remain shrouded in a thick blanket of hydrogen. Some models suggest that a planet 1.5 to 2 times the mass of Earth would have so much extra hydrogen and helium that even with the star's ultraviolet radiation stripping away the atmosphere atom by atom, wouldn't be enough to remove all of the lighter gases. Our planet's own atmosphere is composed of

78% nitrogen, 21% oxygen, and traces of argon and other gases. It contains no hydrogen at all. As with a stronger surface gravity, an atmosphere denser than Earth's doesn't preclude life, but we are still unsure how a civilization may survive on such a world. (Additionally, the atmosphere is the final threshold to space, and a dense atmosphere makes it a challenge to develop further as a spacefaring civilization.)

The surface temperature of a world is an important factor in habitability. Cloud albedo (when clouds reflect sunlight back into space) regulates a planet's atmospheric temperature, so it plays a significant role in whether or not a planet is habitable. If we detect a planet with a high albedo specific to cloud formation (not a world covered in ice), then we'll know it's a wet world, and probably on the warm side. A greenhouse world would eventually result in the total loss of water on the surface of the planet.

Another factor of habitability of a super-Earth is its atmospheric wind speed. The faster a planet rotates, the stronger the average surface wind speed will tend to be. For example, Uranus, a gas giant much larger than super-Earths, has a rotation rate of about 17 hours per day and winds of hundreds of kilometers per hour. All else being equal, super-Earth wind speeds should fortunately be just a bit more than we experience on Earth, so life may still be able to thrive on these worlds.

Super-Earths Are Everywhere!

Super-Earths are extremely common in the galaxy; they are found around nearly every star that we detect orbiting planets. Interestingly, our solar system seems to be the odd one out in that it *doesn't* contain a super-Earth, while most other systems have *at least* one. Since Earth is thought to be on the smaller end of planets capable of supporting life, the law of averages suggests that most alien civilizations reside on these larger super-Earths, if these planets are found to actually be habitable at all.

PUSHING THE BOUNDARIES OF EARTH-LIKE

While there is a significant range in star sizes, the vast majority sit in a single category. As we talked about in the previous chapter, the largest stars (types O, B, and A) will burn themselves out before complex life has a chance to evolve on their orbiting planets.[2] They account for about 1% of all stars. F-, G- and K-types add up to about 10% of all stars. Another 10% are dead stars and other oddities. We are left with nearly 80% of all stars that we are not quite sure can support habitable worlds at all.

This entire 80% rests with M-dwarf stars. They are so plentiful in the Universe that their sheer number alone demands that astronomers rigorously investigate the potential for habitability. There are two reasons they populate every corner of the galaxy: they last a long time and are produced in stellar clouds like weeds in a garden. Much like any objects found on Earth, it's typically easier for nature to produce the smaller variety. The tiniest of M-dwarfs are not all that much larger than the planet Jupiter.

So far M-dwarfs are proving very promising for finding planets around them. The question remains as to whether or not those worlds are habitable. Habitability is dependent more on the size of a planet than on any other factor. The size of all detected planets around M-Dwarfs so far range from gas giants like Jupiter all the way to Mars-sized bodies (and undoubtedly there are smaller exoplanets, we just haven't detected these worlds yet), so size won't be a problem. We have reason to get excited that at least some of those worlds may be habitable.

An M-Dwarf's Younger Years

For the first couple of billion years or so of an M-dwarf's life, the star will go through the same active flaring our Sun did in its youth, except at an increased rate with even more powerful episodes. The intense output of radiation from these flares may severely disrupt the chances for life to thrive on an orbiting world, both because of the radiation itself, and because the star produces more sunspots during this active time.

Sunspots are areas on the surface of a star where the star's magnetic field becomes twisted, producing intense energy that is ready to be unleashed at any moment in the form of powerful flares. The surface of a sunspot is actually much cooler than the material beneath. On stars like our Sun, sunspots can increase its light and heat output – possibly for months at a time. These sunspots can be a significant cause of climate shifts, such as Earth's Maunder Minimum event that lasted from approximately 1635 to 1720 A.D.[3] During this period, as the number of sunspots plummeted, so did the planet's average temperature.

Sunspots on M-dwarf stars function differently; they can be so enormous that their cooler surface areas cause the opposite effect, in that they lower the light and heat output of the star. This reduction in light and heat can be devastating to any life on orbiting planets by orders of magnitude of what Earth experienced during the Maunder Minimum event.

What would it be like if an M-Dwarf star's sunspots blocked out 10, 20, or even 30% of the incoming light for months at a time, dropping temperatures on an orbiting planet not by just a few degrees, but possibly by hundreds of degrees? By comparison, if this dimming happened to Earth, many plants and animals would suffer severe frostbite and eventually – or suddenly – die. The plants and animals that hibernate when winter arrives would need to hibernate for longer periods. If life is possible on such a world, that life may produce amazing new features and abilities to adapt to extreme and unpredictable changes in temperature.

At its most active periods, our Sun spits off as many as twenty flares a day. Young M-dwarfs may flare hundreds of times a day and emit giant flares that temporarily double the star's brightness. Solar flares are bad news for life, especially when the flares are hundreds of times more powerful than those our own planet has ever experienced. Flares can overpower a planet's magnetic field, especially on a smaller world similar in size to Mars where the field is probably going to be weaker. The atmosphere of a small planet could be stripped off by the solar wind. Life may never have the chance to even get started.

A larger planet's atmosphere though could survive an attack by the parent star's flaring activity. A super-Earth with a radius 1.2 to 2 times that of our planet may have an atmosphere that is dense enough for a thin but life-sustaining layer of gas to remain, post-flare. An M-dwarf's more active early years with its constant solar flaring may help clear away some of the lighter gases. The remains of the atmosphere billions of years later may then be earth-like. Then again, the parent star may leave the planet a dry tinder if the flaring activity goes on for too long.

Tidally Locked Worlds

If you look up at our moon on a clear night, and do this repeatedly over the course of several evenings, you will notice that the same side of the moon faces Earth each night. How could this be if everything in the solar system is rotating? Shouldn't the moon be rotating as well? Yes, and it does, but we don't notice it because the rotation is exactly in sync with our planet's own rotation. Like two ballerinas staring into each other's eyes, they are barely aware that it's not the room that's spinning, but they themselves.

The moon's rotation has slowed Earth's rotational rate considerably since the moon first formed. At its formation, our planet used to spin so fast, a day was less than twelve hours long! Over time, the gravitational pull between Earth and the moon, as well as other bodies in the solar system, caused the rotation to slow. While an Earth day (one rotation) has doubled in time, the change has been even more pronounced for the moon because of the difference in size – Earth's gravity influences the moon more strongly. The moon now rotates so slowly, it is in what is called a tidally locked position, rotating only once for every revolution around our planet. This 1:1 orbital resonance is why the moon will always face the planet in the same way.

Any planet that orbits its star at half or less the distance that Earth orbits the Sun will also become tidally locked. The day and night cycle we find natural on Earth would not exist on these worlds. The star would forever shine in the same spot on the near side of the planet with the far side in perpetual darkness (disregarding binary star systems for the sake of simplicity here). Mercury is this way, forever tidally locked with the Sun.

Whether it's huge gas giants or tiny orbiting moons, the tidal lock effect is the same. A tidally locked world then must be detrimental for life... or perhaps not?

Habitability of Tidally Locked Worlds

When planets were first confirmed around M-dwarf stars by the Kepler spacecraft, scientists thought that such worlds would bake on one side and lock up all the water into ice on the other side, making it impossible to foster life. It makes sense at first glance, but it turns out that simulations reveal other factors that may keep the planet from becoming half-snow cone/half-burnt toast. A convection process similar to atmospheric winds might keep water flowing across the planet, regulating the hydrological cycle, as well as moderating global temperatures. Simulations show that as long as there is enough water, ice floating to the day side will repeatedly melt and cycle back via deep ocean currents.

The regulation of temperature and the hydrological cycle will tell us where life can reside on the planet. If life cannot survive on the day or night sides, life may still reside along the narrow band between the two sides, known as the terminator. If life can only exist along the terminator, there could still be plenty of land for life to one day build a civilization, especially if the planet is relatively large. As a civilization advances technologically, it can then make use of the more extreme environments across the rest of the planet, just like we do today in Antarctica, Siberia, and other areas on Earth.

M-dwarf stars radiate more infrared energy than other star types. Plants growing on an orbiting planet will need to evolve ways of capturing energy in infrared. Our Sun produces a lot of energy in multiple wavelengths, so plants don't need to absorb every wavelength; if they did though, the plant would appear black to us. Instead, they reflect the green part of the visible light spectrum. On a planet orbiting an M-Dwarf star, in order to maximize energy absorption, plants will probably need to absorb all wavelengths, and thus are likely to be dark in color. If you ever wore a black shirt on a hot summer day, you have an idea of how much of a difference color can make!

As was mentioned in the opening chapter, even though the interior composition and rotation of a planet is critical in forming of its magnetic field, a tidally locked world is still technically rotating as it revolves around its star. While we have not yet confirmed any tidally locked world having a magnetic field, the speed at which such a world completes a revolution should be enough to generate a magnetic field of sufficient strength (according to some models at least) to protect the atmosphere.

Movie Example – White Dwarf

What a fantastical concept – a world where one half experiences perpetual daylight and the other half experiences perpetual night. How would a civilization survive along the thin terminator that separates the two sides? There are more than a few fantasy and sci-fi stories that explore what a civilization on such a divided world may look like.

In the movie *White Dwarf* (1995), humans find a habitable world tidally locked to its star. Colonies have developed across the planet, on both the day and night sides. The day side is a bustling place with sophisticated technologies; it enjoys perpetual sunshine, beautiful rolling hills and plentiful fields of food. The inhabitants are social and appreciative of the arts. Meanwhile, the night side is a war-torn medieval kingdom; it is blanketed with raging storms and devastating tornadoes, but it benefits from numerous mines full of rare metals that are used for war with the day side. The dark side seems frustrated with its less than ideal landscape, and it is always attempting to stir up trouble for the day side. A massive wall separates the two along the planet's terminator, much like the

Great Wall of China that functioned to keep the Mongolian hordes from invading.

Other examples of films and literature presenting tidally locked worlds include the movie *Star Trek: Nemesis*, a speculative documentary, *What if the Earth STOPS Spinning*, and Isaac Asimov's novel, *Nemesis*, among many other stories.

Just a Little Instability

M-dwarf stars can last for trillions of years. While planets around these stars may therefore have trillions of years for life to develop, continual environmental change will be key for that life to advance into intelligent beings like ourselves. If our past environment had been more stable, there may never have been a reason for humans to evolve into quick-witted and long-legged creatures that could escape a predator, war, or drought. Similarly, higher intelligence may not have evolved at all if there had been no need to migrate to other areas, which forced social interactions with other groups. A peek at Earth's own history suggests that there is a delicate balance between stability and change.

We don't know what the environment on M-dwarf planets is actually like, but if life exists on these worlds at all, it is bound to be interesting and different from our own. What we have learned so far about exoplanets is that they have surprised us – every preconceived notion about what they should be like has been turned upside down.

OTHER, EVEN MORE EXOTIC WORLDS

Habitable Moons

While M-dwarf star systems are the most abundant type in the galaxy, host to strange worlds that call into question the limits of habitability, they are not at the top of the list of the strange and unusual. That spot is reserved for moons around gas giants. These moons are still going to be tidally locked, but to their parent planet, not the star.

While a tidally locked world in itself may not be a problem for life, the smaller size of a moon will have more serious consequences. The reduced gravity of a small moon will make it difficult to maintain a thick enough atmosphere and an active geology with rigorous plate tectonics. Without an atmosphere or active geology, liquid water and a recycling of the atmosphere (to reduce toxic gases and to regulate the greenhouse effect) may not be possible.

There are a few exceptions in our own solar system that defy the expectation that a moon will lack a thick atmosphere and active geology. One very special exception is Saturn's largest moon, Titan, which can hold on to an atmosphere thanks to its cold environment. Titan also had an atmosphere partly because of episodic outgassing from its interior.

Another interesting moon around Saturn is Enceladus, which is suspected to have a massive liquid ocean underneath kilometers of ice. While Enceladus has no atmosphere to keep water from evaporating, the gravitational stresses from Saturn cause the moon to flex enough to drive an active geology underneath the ice crust. The heat produced from the friction process makes it possible for the water to be liquid and to form the ocean beneath.

Saturn's immense magnetic field does a pretty good job at shielding both Titan and Enceladus from direct solar radiation, which would otherwise sweep away any atmosphere they have, halting any evolving life dead in its tracks. The moons have to be close enough to the planet though in order to be sufficiently protected, but not so close as to be affected by radiation coming from the planet's own magnetic field.

An interesting mathematical equation called the Hill Radius suggests that there is a limit to how distant a moon can be to its parent planet without being gravitationally overtaken by other bodies in the system.[4] For example, let's say a gas giant is orbiting where Earth now orbits the Sun, and the moon is now an Earth-sized body. All else being equal, because of the Hill Radius, in order to maintain a stable orbit, an orbiting moon would need to hug the parent planet extremely close, otherwise the Sun and other solar system bodies would start to influence the moon. It would either end up migrating to its own orbit around the star, or be thrown out of the system altogether.

The movie *Avatar* (2009) is a strange example of an exomoon with an abundance of life. From what is currently understood, an exomoon with this much life is not entirely out of the question, but of course there are going to be those pesky physical limitations to consider, like the thin or even nonexistent atmosphere due to very low gravity. Even with these limitations in mind, some scientists think that there might be as many habitable exomoons in the galaxy as there are exoplanets! Examples need to be discovered, otherwise there are simply too many variables to build a reliable model.

Rogue Planets

The configuration of our solar system turns out to be atypical. In fact, it's unique, and so is just about every system we discover with respect to every other. There are some similarities, but they differ wildly. When other star systems are searched for planets, we often find the planets orbiting their parent star extremely closely. These systems sometimes have super-Earths, or lack gas giants. The strangest planetary system so far has been one that appears to have large moon-sized comets careening around the star in elongated orbits, coming as close to the star as Mercury, and then going as far away as Saturn. These eccentric orbits cast doubt on the possibility that any other planets could remain in orbit for long.

Despite the wide variety in system configurations, we can safely assume that a planetary system is a very chaotic place in its early stages of formation. Planetary orbits will shift as the planets jostle for position over the most stable spots in the system. As they gravitationally tug on each other, many will collide, eject others out of the system, or crash straight into their star. After this battle settles down millions of years later, a final stable configuration of orbiting bodies results.

During the formation of our solar system, some planetoids (small bodies with diameters from one to several hundred kilometers) merged to form the planets we see today. Earth was not isolated from these events. As our planet was forming a crust, a Mars-sized body happened to be on just the right angular path to collide, providing enough momentum for the resulting debris to form the moon. The asteroid belt is a remnant would-be planet that did not survive this billiard game.

Planets that get dealt the unlucky fate of being ejected out of the system entirely are going to have a very cold and dark existence, forever drifting between the stars, and perhaps eventually taking its leave of the galaxy altogether. Another star may one day capture the escapee, but the chances of this are very low; an amateur golfer has a greater chance of hitting a hole in one (which happens to be 1 in 12,750).

These rogue planets that meander the Universe may be surprisingly numerous. Some estimates suggest that they are 100,000 times more common than stars, at least in the Milky Way. It's such a fantastically high number, the calculation will give an error on a standard calculator. At such a number, we start to lose sense of just how many rogue planets we're talking about.

What would the surface of such a rogue, aimless world be like? As the planet recedes from the star, temperatures would drop in a matter of hours, just like they do on the night side of Earth, except that temperatures would keep dropping. Depending upon the thickness of the atmosphere, any internal heating processes, and volcanic eruptions, eventually the planet would reach a thermal equilibrium with the cold vacuum of space.

Interestingly, even on a rogue planet there's a chance for life to survive. Take a larger world that is ice-covered. Deep oceanic vents will warm life and provide the energy and nutrients that it needs. Abbot and Switzer, a pair of astrophysicists at the University of Chicago, calculated that a planet 3.5 times the mass of Earth would be warm enough at the core to maintain a liquid ocean beneath an ice crust a few kilometers thick, where life could lurk. They suggest that this ocean could last for as long as five billion years – certainly long enough for life to evolve.

One of these rogue planets is thought to be Nibiru, otherwise known as "Planet X." Conspiracy theorists suggest (without any evidence whatsoever) that it is on a collision course with Earth. Even if a rogue planet were indeed on a path toward the inner solar system, it would not be a concern for us anytime soon. Planets as far as 1,000 AU (Astronomical Unit) should be detectable.[5] At that distance, it would take at least a few decades for it to get close enough to start affecting the orbits of the outer gas giants. While as yet unconfirmed, "Planet 9" is suggested to be several hundred AU out from the Sun. The path it likely takes will never come close to the inner solar system. We are safe from its influence.

A COLLECTIVE EVOLUTION

The Universe is aging in a way that will make it either more conducive to life, or less so, and much of this depends on which elements in the Universe increase and which decrease. We know that life on Earth is carbon-based. It might surprise you to learn that our planet is considered a silicate planet and not a carbon one. While there's obviously enough carbon to form life, the vast majority of material is silicate-based. Mountains and sandy beaches are composed of silicates.

One might assume that a carbon-rich planet would make an even better home for life than Earth, as it has more of the very element we're made of – carbon. But this is not necessarily the case. For instance, the carbon-to-oxygen ratio of Earth provides a low enough carbon part to allow water vapor to form; without vaporization, the hydrological cycle is thwarted. Higher levels of carbon may also produce a hazy atmosphere of natural pollution, complicating plant growth and, by extension, the evolution of any animals. A highly carbon-rich planet may resemble Saturn's moon Titan, which has a haze of hydrocarbons in its atmosphere. For our planet, this would be a pollutant, but a different kind of life may still be able to evolve in such a haze.

Currently the carbon-to-oxygen ratio in high-metallicity stars and their planets is approximately 0.4 to 1.0. While our solar system is set in its ratio of these two elements, this won't be the case for future systems yet to be born. Every generation of planets will tilt towards being more carbon-based, and that may have significant consequences for how abundant life can get. Whether those consequences are positive or negative, we don't know – we have to discover other life-bearing worlds to find out.

Though alien life may be recognizable in some ways, other ways are just as likely going to be completely unique. How fantastically unique can life get? We really have no idea yet - all we know is that alien life will evolve according to the laws of nature and physics. Imagine creatures larger than whales flying high in the sky of a super-Earth, brushing clouds of methane, or perhaps an entire planet's ecosystem interconnected so much that it is essentially one living, breathing, and thinking organism. Or how about a creature with two independent brains?

HABITABILITY ZONE SAMPLE

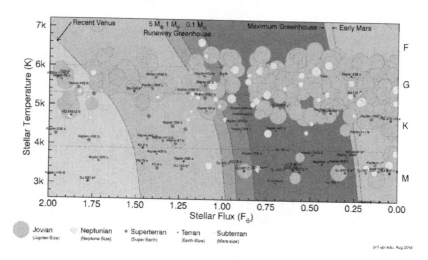

The above chart depicts the range of habitability for known exoplanets.[8] The circles are representing the size of a planet. The

Stellar Flux axis notes the energy output of the planet's parent star, and the Stellar Temperature axis is showing the star's temperature. Planets depicted to the lower left are thus closer to their parent stars. The most important point to note are the bands. The darkest band depicts just the right distance from a star where a planet could have liquid water on its surface, and thus be habitable. The percentage of greenhouse gases in a planet's atmosphere will shift its placement along the graph.

The Universe, at least our corner of it in the Milky Way, seems to have produced exoplanets in abundance.[6] As detection techniques improve, even more exoplanets will be discovered. While gas giants and planets closer to their stars currently dominate the overall count, eventually they should be in the minority, since these two types of planets are easier to detect. The discovery of a distant Pluto-sized exoplanet, for instance, is going to come far later than a system's inner planets.

Upcoming telescopes present a good chance for us to find out the true nature of super-Earths and other exotic worlds.[5] We should soon be able to confirm the existence of plate tectonics and any magnetic field by observing the effects shifting plates and an active magnetic field have on a planet's atmosphere. From what detection techniques have been developed so far, it's likely the secrets these worlds keep will eventually be revealed in the next couple of decades. We will then be able to apply learned techniques to analyze the atmospheres of even smaller worlds, perhaps one day actually finding a truly habitable one![7]

CHAPTER 8: THE SCALE OF THINGS

"Space is big. Really big. You just won't believe how vastly, hugely, mind-bogglingly big it is. I mean you may think it's a long walk down the road to the chemist, but that's just peanuts to space."- Douglas Adams – The Hitchhiker's Guide to the Galaxy

The practically infinite scales of space and time are nearly incomprehensible. Time spans from the billions of years the Universe has existed to the fraction of a second that it takes to split an atom in a nuclear reactor, and time includes the middle ground of a still brief but rich span of an average human life. Scales in distance are just as stupendous. The immensity of space, both in the void between galaxies and the void between atoms, is difficult to visualize.

Experiencing cosmic scales of space and time directly may not be possible, but we can still use comparisons to better understand them. For instance, if the Sun were the size of New York City, then the nearest star, Alpha Centauri A, would be as far away as San Francisco. Looking at it another way, scaling down Earth to the size of a grain of sand, the Sun would be about the size of a grape and one meter away from the grain of sand. Alpha Centauri A would be more than 125 kilometers away!

Whether it's billions of years or billions of ants on an ant hill, we're still counting using numbers that did not exist in the minds of humans just a few millennia ago.

SCALING EXISTENCE[1]

Amount	Scale	Example
Million	10^6	A million seconds is 11.57 days
Billion	10^9	A billion hours ago, humans were just entering the Stone Age
Trillion	10^{12}	About how many stars there are in the Andromeda Galaxy
Quadrillion	10^{15}	A stack of papers reaching to the moon and back, 14 times over
Googol	10^{100}	There are not enough atoms in the Universe to equal a Googol!
Multiverse	10^{500}	Possible number of dimensional shapes

Counting Them All Up

If you hold a pen to the night sky, the tip of that pen would cover an area the size of what is famously known as the Hubble Deep Field (HDF). The HDF is a tiny fraction of the sky that at first looks completely devoid of stars from telescopes with less power than Hubble. The HDF is anything but empty when one takes a closer look. What Hubble found in that pen point is more than 10,000 galaxies. Because the Universe is homogenous, i.e. unvarying across the largest of scales, every other area of a pen point in the sky also contains as many galaxies.[2] Total all of those galaxies up and you get at least 160 billion in the Universe![3]

How do we comprehend the sheer scale of billions of galaxies and trillions of stars? We could start with an analogy and say that there are more stars than there are grains of sand on all the beaches on Earth. That's hundreds of thousands of grains of sand in a single handful, and trillions of handfuls worth. We can extend this analogy into the seemingly infinite to include the hundreds of trillions of planets orbiting the Universe's stars.

Even more out of this world are the number of atoms in the Universe. Because math is so consistent and accurate regardless of scale, we can count how many atoms there are, within a reasonable margin of error. The mass of the different star types can be measured from the atomic weight, multiply that by the amount of stars of that type in the galaxy, calculating in other massive bodies like supermassive black holes, and then how much influence galaxies have gravitationally upon other nearby galaxies. We can get a rough estimate of the total mass of the Universe through all of these gravitational interactions. The actual calculation is of course more complicated than this single paragraph suggests, but it is still possible to estimate.

That number ends up being somewhere between 10^{78} and 10^{82} atoms. Keep in mind that here we mean the *observable* Universe, which extends outward 46 billion light years in any direction.

For a counting example closer to home, an elderly Chinese man named Zhang Ming-hua has lived his entire life along the Great

Wall of China. Since he was a teen, Mr. Zhang has worked on a rather extraordinary project for the Red Guard. His assigned task is to count every single brick in the wall. The wall spans 21,196 kilometers, or 13,170 miles. So far Mr. Zhang has counted the bricks in about 20,500 kilometers of the wall. Mr. Zhang estimates that he has a couple of years left and will have counted approximately 10 billion bricks upon completion.

The Largest and Smallest

We can visualize everyday objects like a school bus, an office building, a large lake surrounded by forest, and even vast mountain ranges that span from one horizon to the next. Objects much larger than a mountain range and we start to have difficulty comprehending how large they really are; we end up needing to use vague and clumsy comparisons and mathematical language. Even our great imaginations quickly fail us once we get much beyond the scale of objects like planets and stars.

For the purpose of comparing the largest to the smallest, let's start with the extreme vastness of the entire Universe. If the Universe were the size of Earth, by extension, Earth in our model would be about the size of an atom, and the solar system would only be the size of a dust particle. The Milky Way galaxy, which is amongst the largest of objects in the Universe, would only be the size of a small house.

Now let's reverse the model and try to grasp the very smallest of objects. While we can imagine forever zooming in, it turns out there actually is a limit. That limit is the Planck Length and is 1.6×10^{-35} meters, or 0.000000000000000000000000000000000016 meters. Numbers this small are impossible for the human mind to comprehend, so we must once again use analogies to grasp the truly infinitesimally small scale of the Planck Length. If the smallest dust particle the human eye could see where scaled up to the size of the Universe, the Planck Length would then be about the same size as the original particle, relative to our model's expanded Universe-sized particle.

When I was a kid, I used to imagine entire universes somehow embedded in every atom of my body. The scales of space and time had no boundaries for my imagination. If we find out one day that space and time are infinite, who's to say that at the Planck Length it ends there? Perhaps if one could zoom in far enough, they would get drawn into an entirely new universe? What if the Universe around us is at the Planck Length of someone else's universe?

EVOLUTION AND DEEP TIME

Deep time is a term used to describe the astonishingly vast period of time over which geological changes, like the moving of continents and the evolution of life, occurs. When we ask the question "what time is it?" we are usually referring to the hour and minutes of a particular day. If you were to ask the question "what deep time is it?" your response might be the geological time period we live in, which is a relatively new epoch called the Holocene, spanning from about 12,000 years ago to the present.

As long as 12,000 years may sound, it pales in comparison to the lifetime of our species. Take 12,000, multiply by 20 and you get an estimate of how long Homo sapiens has existed, which is about 200,000 years, or enough time for a few ice ages and several smaller glaciations to occur. Multiply 200,000 by 9.5 to get 1.9 million years ago and you are in the time that Homo erectus split off from an ancestor that resembles today's great apes. Two million years is enough time for over 8,000 United States to develop from its inception to the present state.

If we compress all of deep time down to a 24-hour clock, starting at the beginning of Earth's formation 4.54 billion years ago, and work our way forward to the dawn of man, we can put deep time into a framework that can be better understood. Each hour represents about 189 million years. Each minute, 3.15 million years.[4]

Starting at 00:00 hours until about 04:00, the planet cools and a crust layer forms. Meteors continually rain down from the sky with a still coalescing moon overhead. Temperatures are into the hundreds of degrees Celsius. Other than a chaotic and hot solar system filled with dozens of planet-sized objects occasionally colliding with each other, things are not very exciting on newly formed Earth. It is at 04:00 that the continents start to form.

From about 04:00 to 09:00, single-celled bacteria are all that evolves deep in the oceans, emitting oxygen as a byproduct. The oceans absorb the oxygen over a billion years, until it reaches a tipping point when oxygen begins to fill the atmosphere, around 2.5 billion years ago, at 12:00 on our clock. Known as the Great Oxygenation Event, this period in Earth's history is one of the most important evolutionary events, as it opened the door to the evolution of multicellular life, and eventually billions of complex species.

Once enough oxygen builds up in the atmosphere, the ozone layer forms. Ozone is made up of three oxygen atoms (O_3), also known as trioxygen, and it blocks the UV radiation from the Sun that is harmful to life. Water also blocks this harmful radiation, and is why life before the Great Oxygenation Event only inhabited the oceans. Now life can move out of the water and onto land. All life still inhabits the water until 18:40 when the first simple plant life manages to evolve on land, suspected to be in the form of blue-green algae (Cyanobacteria). Vascular plants appear at about 21:36, insects at 21:52, and dinosaurs at 22:47.

At 23:40 the dinosaurs die out. The entire age of the largest land animals to ever roam Earth, more than 160 million years of evolutionary history, amounts to less than an hour on our clock. The first significant deposits of carbon and other organic compounds form at this time as well, and they are the deposits we use today as oil and natural gas to power our cars and homes.

At 23:43 the first large mammals appear and dominate the globe, filling a niche the dinosaurs left just a few minutes earlier. At 23:59 and 12 seconds the genus Homo evolves. Finally, at 23:59 and 56 seconds, modern humans, Homo sapiens, appear. They make a grand entrance with only a few seconds left on our clock before it strikes 24:00.

We considered deep time in an example from our past, but what about the future? Star birth and planet formation will continue for trillions of years more. There is a lot of time for the Universe to expand and evolve – much more than the 13.8 billion years it has existed thus far. So the deep time of the future lends great hope for the evolution of intelligent life.

Another Case for M-Dwarfs

Complex multicellular organisms have existed for a few hundred million years, which is a miniscule fraction of the time available for life to develop around M-dwarf stars, which can exist for a trillion years or more. To grasp what kind of opportunity it would be for life to have a trillion years to evolve, consider that a trillion years is about 73 times that of the Universe's current age of 13.8 billion years. A trillion years is also 100 times the lifespan of our Sun, and 5 million times the total number of years humans have existed. You could watch Earth be born, witness life evolve, observe the sliver of time during which humans build great civilizations many times over, eventually see the aging Sun boil away the planet's oceans and evaporate its atmosphere, and then repeat the entire spectacle at least 185 times before the last M-dwarf stars finally burn themselves out.

If, on average, it takes at least a billion years for a planet to become habitable, and another couple of billion more for life to evolve into something capable of intelligence, M-dwarf planets should have plenty of opportunity to evolve the kind of life we are looking for in our telescopes.

Let's assume for a moment that M-dwarf systems never start out with a planet in their habitable zone, but instead the planet resides far out in the icy suburbs beyond the Snow Line, like Jupiter or Saturn does in our solar system. This does not pose a problem for the development of life because, over deep time, a fraction of systems will have planets that slowly migrate into a habitable orbit.

Even when an M-dwarf star is young and emits tons of radiation that strips away much of an orbiting planet's atmosphere, life just needs a thin protective layer remaining, a significant percentage of surface water, and an initial set of chemical conditions to get started. If life can get started at this early stage in the star's stage of development, then deep time alone will favor life evolving into more interesting and hardy forms later on.

For instance, there could be countless resets on M-dwarf worlds where life is nearly destroyed and rebuilds back up again into something complex. Life on Earth has been disrupted, or almost set back to single-celled existence, numerous times since multicellular life evolved 500 million years ago. Every few million years there is a large asteroid strike, supervolcano eruption, or some other disaster. Earth recovers thanks to deep time. It only has about 900 million years left, though, for evolution to play out, before the Sun toasts our planet. The clock is ticking!

In the Blink of a Civilization

We know that civilizations have come and gone throughout history. It's easier to grasp how quickly some of them have vanished over time, but difficult to appreciate how some have survived for centuries, inevitably changing their culture along the way. Among them are the Chinese, Egyptians, Israelites, Japanese, and Mongols. While the descendants of these ancient people thrive in the same lands today, they differ enough from their ancestors so as to be classified as a separate civilization.

About 5,000 years ago in 3100 B.C., civilization in ancient Egypt began. In case 5,000 years doesn't seem like a lot of time, let's put that number into perspective. The average lifespan of an individual back in 5000 B.C. was about 25 years, birthed by parents at roughly the age of 14-20. Over the course of this 5,000 years, on average about 200 generations lived.

Many people can trace their lineage as far back as perhaps a few hundred years (some can go back much further). Keep in mind that this represents just a handful of generations. Take the family tree you see and multiply the count of persons by a hundred. Now you are closer to the amount of ancestors you've had over the last 5,000 years!

Go back far enough and eventually your lineage will converge with mine and everyone else's. This convergence point between lineages is called the MRCA, or Most Recent Common Ancestor.[5] The convergence occurs because not every branch of the family tree survives, and so eventually there comes a point in the past when a single individual will be the MRCA for the entire population that follows in the future.

The MRCA of today's human population dates back to around 65,000-75,000 years ago. If there had been no isolated populations, separated by continents, on remote islands, and other obstacles that prevented breeding between groups, then the MRCA would have lived as recently as 3,000 to 5,000 years ago.

Humanity's Expanding Waistline

Some estimates suggest that the entire human population that has ever walked on Earth amounts to about 108 billion. This is quite generalized, given the time scales involved, and that there is no way that we can establish when one species split off from another with any real precision. Our current global civilization stands at almost 7.5 billion people. This is 7% of the *entire* population that has ever lived!

For nearly the entire span of human history, worldwide population remained much less than a billion persons, and as low as just a few thousand during extreme disasters. After thousands of years of flat growth, birth rates rose sharply during the 18th century, as humans began prospering like never before. World population hit one billion around the year 1800. The exponential population growth we still see today was underway. The steady increase in population provided enough manpower to develop technologies and farming innovations that furthered growth. No longer was the population at the mercy of disease or famine, thanks to scientific progress. From about 1804 to 1927, global population doubled from 1 to 2 billion. The latest billion persons this century were born in just the previous 10 years.

Let's take a fresh look at the population chart from Chapter 4:

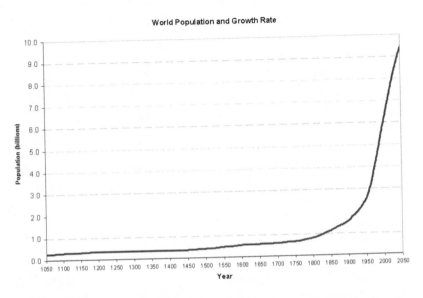

The next billion persons is projected to be born in even less time, though there are some signs that growth will plateau later this century. The next decrease in population will not be because of famine, disease, and war (although any of these could happen, too), but because the cost and time required to raise children, and the need to have children around the farm for work, will become unnecessary or even prohibitive for many people.

DEEP DISTANCE – From Point A to Point B

The Oregon Trail was a popular video game when computers were just becoming available to the public in the 1970s. The most successful version of the game later arrived in schools on the Apple II in 1985. The game helped to raise awareness of the real Oregon Trail in the United States which spanned from the Midwestern state of Missouri to Willamette Valley in Oregon, about 3,500 kilometers (2,200 miles). Development of the Oregon Trail began around 1811; a half-century went by before it was completed and used regularly.

At its peak use in the mid-1800s, more than 412,000 settlers, traders, miners, and others used the trail to access rich farmlands and growing towns along the Northwest coast. Traversing the trail could take as long as 170 days, especially when traveling with children and heavy cargo. If you were a pioneer exploring the Wild West for farmland, the trail could be traversed in as little as 120 days. Today the journey is much safer and would take a few weeks by bicycle, three days by car, and a brisk (though still arguably just as uncomfortable) four hours by plane.

The trek was extremely dangerous, especially along the many rivers that wagon parties had to cross. Wheel axles broke, people fell off and were dragged under current, and frostbite often set in. Illness was rife. Food had to be hunted and eaten on the spot before it spoiled. Threats from Native Americans, bandits, or even other wagon parties posed great challenges as well.

Initial expeditions and journeys into new lands are nearly always full of peril, especially when the distances are long. As we advance in our knowledge of the risks and ways to mitigate them, not only is a safer path created for others to follow, but we also speed up the voyage. Today, flying is the safest and the fastest way to get from Missouri to Oregon. Every technological advance humanity has made over the last two centuries has brought us one step closer to that four-hour luxury we enjoy today, the elbow-to-elbow flying sardine can.

Outer Space - A Vast New Frontier

Outer space provides a limitless new frontier for exploration. The moon is the first stop on this amazing journey. The moon is on average about 384,400 kilometers from Earth, or about 110 times the distance of the Oregon Trail. Apollo 11 reached lunar orbit in 51 hours and 49 minutes.

Three days is just a long weekend, which doesn't sound so bad for traveling to such a cool place as the moon, until you scale distances up. With current space travel technologies, it would take about ten years to get to the furthest planet in the solar system – Neptune. The shortest distance between Earth and Neptune is 4.3 billion kilometers, which is a mind-boggling 1.2 million times longer than the Oregon Trail. (When Earth and Neptune are on opposite sides of the Sun, the distance between them is 4.7 billion kilometers.) Without rest and at the modest pace of 25 kilometers per hour (wagon speed), it would take a minimum of 19,635 years to reach Neptune.

We are making rapid progress in permanently escaping Earth's gravity and colonizing at least the moon. The former Soviet Union, the United States, China, Japan, The European Space Agency and India have sent unmanned spacecraft to the moon, and about 20 of them touched down on the lunar surface. The United States is the only country that has sent manned spacecraft to the moon – 6 times. A total of 12 men have walked on the moon.

The era of moon visits kicked off in 1959 when the Soviet Union launched the Luna 2 spacecraft. Luna 2, or Lunik 2, was the first object from Earth to touch down on the surface of another body in space. It gathered basic scientific information about the lunar surface, such as radiation levels, seismic activity, magnetization, etc.

After a decade of sending machines to the moon, in 1969 the United States spaceflight Apollo 11 successfully sent the first humans. When Neil Armstrong and Edwin "Buzz" Aldrin stepped foot on the surface of the moon, it was the first human footstep on any body other than our home planet. Armstrong and Aldrin spent about 135 minutes exploring the lunar landscape, observing and collecting lunar material. The flag of the United States was also firmly planted in the ground with the flag cloth unmoving for lack of an atmospheric wind.

Neil Armstrong's famous words moved many to tears, as 530 million Earthlings watched him on television: "That's one small step for a man, one giant leap for mankind."[6]

To this day, most of the rovers and other objects sent to the moon still rest on the lunar surface in nearly the same condition as the day they arrived, principally due to the moon lacking a dense enough atmosphere to weather them away. The only significant change would be the flags and other markings on the rovers. Over the years in direct sunlight, these markings would be bleached white from the Sun's ultraviolet radiation. Perhaps this is a fitting outcome, since the moon is loved by all of humankind, not just the particular countries able to plant their flag upon its surface.

Travel distances to the moon and other planets in the solar system is trivial though when compared to interstellar distances. The vastness of space is unfathomably enormous. Let us nevertheless try to fathom it.

Light travels at a speed of 299,792,458 meters per second. That's 186,000 miles per second, 671 million miles per hour, or 1,080 million kilometers per hour. This is fast enough to go around Earth 7.5 times in a single second. Now, if light can go that far in a second, imagine how far it can travel in a year. The answer is an astounding 9,460,730,472,580,800 meters, 9.461 trillion kilometers, or 5.878625 trillion miles. A light year is almost 10 trillion kilometers long!

There is debate about the length of the Milky Way, with estimates falling between 100,000 light years and 180,000 light years. Let's use the conservative estimate. Multiply 9.5 trillion kilometers (the distance of one light year) by 100,000 light years (the shortest length our galaxy is likely to be) and we get over 9,000,000,000,000,000,000,000 kilometers!

With a galaxy at least this vast, chances are that Earth is not the only place where life exists. But with distances between worlds extreme, making contact with other civilizations seems a monumental challenge, to say the least – even if we could travel at the speed of light.

Richard Garriott, famous video game developer and the first private astronaut, had this to say to me about his views on the vastness of outer space:

"I was 47 years old before I finally made a complete circumnavigation of our home planet. On October 12, 2008, I left Earth aboard a Russian Soyuz and lived aboard it and the

International Space Station for 12 days. Before this trip, I felt the Earth was a huge place filled with innumerable people, places and things. However, after seeing the Earth from space I experienced "The Overview Effect" as many spacefarers do, and the Earth become suddenly finite and small. In contrast, even with the velocity of our fastest spacecraft, we have barely reached outside our own solar system since the 1970's. Even at the speed of light the nearest stars would be decades away. The nearest galaxies, so unimaginatively far, that billions of years at light speed would be required to traverse such distances. So, even having traveled faster and farther than most other humans, even with the new sense of scale garnered from such experiences. It seems to have only deepened my sense of awe about the true scale of the visible Universe, much less what infinite realities may exist beyond it." - Richard Garriott (Soyuz TMA 13 / 1st Second Generation Astronaut / Video Game Developer)

The Universe is so vast, and expanding at an ever faster rate, that light at one end will never be able to reach the other end. It would take roughly 93 billion years for light to travel this distance. This fact becomes even more amazing when you consider that the Universe is only 13.8 billion years old, and has trillions of years left to expand ever larger.

Eventually, billions of years from now, all that may be visible from Earth will be the immediate galaxy surrounding us, for everything else will have receded away as the Universe continually increases its expansion. All the other galaxies and any of their inhabitants will view themselves in the same way. We are fortunate to appear while the Universe is so young and revealing to us!

Theoretical Travel Possibilities

The proximity of the moon provides us with a perfect opportunity to test many spaceflight technologies. Once a permanent presence on the moon is established, we can use its resources to reach even further into space. There is a lot to like about the moon, such as its valuable helium-3. This version of helium does not occur in significant quantities on Earth, and it is quite valuable for use in nuclear fusion research. If we could harvest the moon's helium-3, spacecraft could be powered locally. Eventually the spacecraft could be built on the moon and sent to other places in the solar system, such as onward to Mars.

A number of experimental propulsion technologies are currently being tested in labs, such as advanced ion, nuclear fusion, antimatter, and other futuristic engine designs that use exotic forms of energy. Currently the most powerful would be an antimatter drive. Antimatter is a form of matter with opposite charge. If it comes into contact with matter, both forms are completely destroyed and pure energy is produced. All forms of movement, or work, require energy, but all end up losing some of that energy as heat. Efficiency is never 100%, except when matter and antimatter collide. If this process can be controlled, spacecraft speeds could reach 20% the speed of light or more. At this speed, we could get an unmanned spacecraft to Mars in about 15 minutes!

If an antimatter drive is possible, we could theoretically get to the nearest solar system, Proxima Centauri, in about 40 years, if we did not slow down to enter the orbit of any of its planets.[7]

An Alcubierre drive is a concept proposed by Miguel Alcubierre and is similar to the warp drive described in numerous science fiction stories, namely *Star Trek*.[8] The drive would use what's called negative energy to create a space bubble around the ship. This protective bubble would squeeze space in front of the ship and expand it in the back, thus propelling the ship forward. Theoretically, the speeds achieved could be many times that of light. You are still not violating the light speed limit because you are not moving *in* space, but around it. It's a strange but important distinction. The Alcubierre drive lies within the realm of the theoretical, but apparently it doesn't violate any laws of physics, and is mathematically consistent throughout its design.

Using the theoretical Alcubierre drive, we could get to Proxima Centauri in a matter of just a few days. Journeying across the galaxy would still take many years, but it would be in the realm of an individual's lifetime.

Then there are wormholes, which are also mathematically possible, but they are just as deep in theoretical territory as the Alcubierre drive. As fascinating as these concepts are to suggest, we should probably remain skeptical about what the Universe both contains and allows us to use. We just don't know where the actual boundaries lie… yet.

THE GREAT COSMIC OCEAN

The Universe has existed for 13.8 billion years, with the Milky Way nearly as old at 13.2 billion years. Yet our solar system is *only* 4.6 billion years old. Life had billions of years to evolve on other worlds in the galaxy before Earth was a newly formed molten ball of rock. Many scientists believe that our galaxy, at its current age, is very young; it will live on for many, many more billions of years. Peter Behroozi, lead researcher at the Space Telescope Science Institute, thinks that because there is so much gas and dust left to create new star systems, most habitable planets and their parent stars have yet to be born.

Long after our Sun dies out 5.5 billion years from now, new stars will be born that will last for billions, perhaps trillions, of years. When the Milky Way and Andromeda galaxies merge, bouts of star formation will occur. (This new galaxy is being called "Milkomeda" – although I prefer "Andromilky.") Stars trillions of years from now will still be born deep in the bowels of this new galaxy's final star factories.

Just like how stars are born, and then die, the Universe itself has an age limit. About 10^{40} years after the last stars burn out, which is trillions upon trillions of years from now, the very atomic structures that matter is made up of–will fall apart and cease to exist. The only thing that will remain is a uniformly thin form of energy – no stars, no planets, no dust, and no gas. The Universe will settle into such a low-energy state that, for all intents and purposes, it will be dead. This ultimate end to existence is a theory known as the "heat death" of the Universe, and the final chapter of our cosmic story… as far as we know.

CHAPTER 9: IS ANYBODY OUT THERE?

"In very different ways, the possibility that the Universe is teeming with life, and the opposite possibility that we are alone, are equally exciting. Either way, the urge to know more about the Universe seems to me irresistible, and I cannot imagine that anybody of truly poetic sensibility could disagree."
- Richard Dawkins

Despite all that we have come to know about the Universe, there is still so much left to understand about the probability of life elsewhere. There are many unknowns yet to explore, and surely just as many that we can't even imagine, but we can start with what we know about Earth and the solar system to hypothesize. The gaps in our knowledge are being filled in with new data all the time, but scientists admit that there is a lot of uncertainty about what we might yet find.

Other planetary systems discovered so far indicate that our solar system is not the typical configuration of planets. In fact, we have thus far detected no two similar systems. The Milky Way contains a virtual zoo of various sizes, orbits, and compositions. Some systems have massive Jupiter-sized exoplanets hugging their parent stars, while many have multiple super-Earths. Others have no asteroid belt at all, and at least one system we know of has a massive asteroid belt many times greater in width than our own. There are even some with four or more orbiting planets that could fit within the orbit of Mercury!

Partly because planetary systems are so diverse and not of the expected norm, detecting life on these other worlds is one of the most challenging things humans have ever set out to accomplish. Excitingly, we are just now developing the technology to detect the gases in the atmospheres of exoplanets, which will tell us a lot about any life on these worlds. Certain gases like oxygen and methane, especially in combination, may indicate the presence of life. Future generations of telescopes will greatly improve our ability to make more detailed observations.

Detecting life's signatures on these worlds may be as far as our technology will ever be able to provide. Because of the vast distances involved, we may never know how alien life differs from that on Earth. They may have a host of unique senses that don't even exist on our planet. Life must conform to the laws of physics and chemistry regardless of where it comes from though, so some features are bound to be similar to creatures we visit at our local zoo, while others are just as likely to be more different than our wildest science fiction stories can craft.

THE DRAKE EQUATION

Frank Drake is an American astronomer and astrophysicist. He was an early prodigy in the sciences, experimenting at school with electronics and chemistry before most of his classmates cared about such subjects. It was not long before he started to ask the question, "What are the chances of there being intelligent life elsewhere in the galaxy?" In 1960, he got the opportunity to try and answer that question with Project Ozma, which was the first attempt at detecting an alien signal.[2] The project was an important precursor to what would be called the Drake Equation.

Later in 1960, Drake started working on the Drake Equation, though not quite in the way one would expect; it was more of a curiosity at the time. Little did he know that it would become the de facto standard for calculating the theoretical chance of life and civilizations elsewhere.

The Drake Equation is not as intimidating as you might expect. There are no charts or endless pages of formulae. Instead, the entire equation is composed of just seven factors on a single line. You get a result by multiplying all of the factors. That's it! We'll start with the best understood factors and work our way to those yet to be quantified.

Analyzing Drake's Equation

$N = R_* \times f_p \times n_e \times f_l \times f_i \times f_c \times L$

N - Number of technological civilizations in the Milky Way

R_* - Rate of star formation per year

f_p - Fraction of those stars that have planets

n_e - Number of habitable planets, per star that has planets[1]

f_l - Fraction of those planets that go on to develop life

f_i - Fraction of life-bearing planets that develop intelligent life

f_c - Fraction of those intelligent life forms that emit signals into space

L - Length of time the civilization continues to emit signals

R_*

R_* is the first value and the part of the equation that is best understood by astrophysicists. R_* is the average number of new stars born in the galaxy each year. Drake originally estimated this number to be 1, but later he suggested it could be as high as 10. Today NASA says this turns out to be a healthy... 7. That may not sound like a lot, but keep in mind that the Milky Way has existed for about 13.2 billion years, and star formation was much more active in the first few billion years after the galaxy formed.

f_p

f_p is the second factor and one we are just now able to calculate fairly accurately, thanks to observations made using the Kepler Space Telescope. This factor is the number of the existing stars that have planets. It was once thought to be anywhere from 10% to as high as a half of all stars. Astronomers used to believe that planets could not form around binary stars (star systems that have more than one star), which comprise nearly 50% of all stars in the Milky Way. Since exoplanets have been discovered, it has been found that nearly all stars, binary pairs or not, have at least one planet orbiting them.

n_e

The third factor, n_e, considers how many planets are capable of supporting life. To support life, a planet must orbit within the star's habitable zone. It also must orbit a star that lasts long enough for life to have time to develop. Additionally, the planet needs to be of the right size and composition. All of these pieces of the puzzle are jammed into n_e, which narrows our final estimate significantly. We're also neck-deep into the realm of uncertainty here, but not yet drowning in ignorance, thanks again to the Kepler Space Telescope and other tools. With more than 3,000 planets confirmed by KST, and thousands more candidate signatures that are being confirmed as planets, it's beginning to look like there is a significant number of habitable worlds out there!

f_l

Now we get to factor f_l – the fraction of habitable planets on which life actually does evolve. This is where we cross from uncertain to the completely unknown. Scientists have spent years refining estimates of how often this might occur, but we have no actualities. Thus far, we only know of one confirmed habitable planet with life on it – Earth. Drake thought the fraction was 1, or 100% of habitable planets would develop life, though this is unlikely given the limits we see within our own solar system.

Even if the chance of a habitable planet hosting life is at 100%, planets that only can host a few microbes and never anything more interesting do not titillate us; we're interested in whether a planet has a chance of spawning an intelligent civilization. Mars may very well be able to host microbes, but clearly not anything that can walk and talk.

f_i

f_i takes into account how many life-bearing planets will go on to develop intelligent life. We are back to having a bit more understanding, based on evolution and the laws of nature as they have played out on Earth. The rarity of intelligence on our own planet might suggest that evolving an intelligent brain is unlikely, but Drake was very confident that if there was complex life on a planet, an intelligent creature would evolve, if given enough time. I tend to agree with him here.

f_c

Once there are thinking beings that can manipulate their environment, they will likely gain the ability one day to create a radio antenna and say hello to the rest of the Universe. Exceptions would include water worlds or some other physically, but not intellectually, restrictive environment.

There could be billions of planets hosting life with millions of civilizations huddling around campfires, or swimming on their water worlds, occasionally peering up at the sky in wonder. That doesn't do us any good in learning about them. To learn about far-off worlds, we need them to either develop signal technology, and to emit detectable signals, or to alter their planets' atmosphere sufficiently for us to detect industrial pollutants. This would help us gauge what technologies they have, and by extension how advanced they are in comparison to us.

L

L is the amount of time a civilization's ability to send signals into space lasts. As outlined in Chapter 5: A House of Cards, numerous disaster scenarios can befall a civilization during its development. Nuclear war has come close to destroying our own world a number of times, and we've only had nuclear weapons for less than a century. Drake suggests a rather wide range for factor L: on the low end, a healthy 1,000 years and on the upper end, a fantastically generous 100 million years.

Here are the values that Drake used:

$R_* = 1$/year (1 stars formed per year; this was regarded as conservative)

$f_p = 0.2\text{-}0.5$ (one fifth to one half of all stars formed will have planets)

$n_e = 1\text{-}5$ (stars with planets will have 1 to 5 planets capable of life)

$f_l = 1$ (100% of these planets will develop life)

$f_i = 1$ (100% of which will develop intelligent life)

$f_c = 0.1\text{-}0.2$ (10-20% of which will be able to communicate)

$L = 1000\text{-}100,000,000$ years (ability to emit signals will last this long)

N

Let's multiply Drake's factors and see what the product, N, is.

$N = R_* \times f_p \times n_e \times f_l \times f_i \times f_c \times L$

The equation with Drake's lowest values:

$N = 1 \times 0.2 \times 1 \times 1 \times 1 \times 0.1 \times 1,000$
$N = 20$

The equation with Drake's highest values:

$N = 1 \times 0.5 \times 5 \times 1 \times 1 \times 0.2 \times 100,000,000$
$N = 50,000,000$

According to Drake's estimates, there are between 20 and 50 million intelligent communicative civilizations. Try the equation out with your own numbers!

DETECTING ANOTHER CIVILIZATION

Sending a Signal

Sending out radio transmissions is one of the easiest ways to send information long-distance. Advantages of radio signals include that they are detectable far beyond conventional human senses, if the transmitter and receiver are both powerful enough. Radio transmissions can be sent in many directions at once, reaching thousands of stars. Also, it should be clear to intelligent civilizations that the radio signal is from an intelligent source.

Elements on the periodic table have their own associated electromagnetic absorption frequencies. These frequencies will be consistent for every element, wherever they are found throughout the Universe. Hydrogen is the most common element and it has a radio frequency of 1420.4~ MHz. Using the frequency of the most common element in the Universe in our radio transmissions will greatly increase the chances for a listening civilization to pick up the signal amongst the many other frequencies available.

Deciding what we ought to send to an alien civilization intrigues astronomers. Many think we should specify how far humanity has progressed scientifically and technologically; a receiving civilization will be able to understand a lot more about us if they have a grasp of how far along the technological ladder we are. As explained in Chapter 4: The Engine of Modern Civilization, because it is difficult to advance technologically, in that it requires intelligence and cooperation, our state of advancement will indicate that we have some altruistic goals that would appeal to any aliens interested in contact.

Showing our technological prowess can be done by sending mathematical proofs, like Fermat's Last Theorem, chemical formulae, such as complex man-made ones, or our discoveries in physics, like the Higgs boson. All of these would be encoded in the radio wave sent out into deep space, much like how a complex message can be sent with Morse code.

There are panels of scientists at conferences each year that discuss what kind of message makes the most sense to send. One prominent event is Exoplanets, Biosignatures, & Instruments (EBI). At the conference they discuss whether we should send only information about the scientific features of our culture, or information about everything that makes us who we are as humans. The concern with sending information about our entire culture, including our fictional and artistic creations, is that the aliens may misinterpret the message. They may not understand where our science ends and our art begins; they may not know our fact from our fiction.

Being detected by an alien civilization is actually a worry in the eyes of some scientists, like Stephen Hawking, who suggest that our signals could be received not by a benevolent race, but by an aggressive one, like the ones portrayed in the movies *Independence Day, War of the Worlds,* and the humorous *Mars Attacks!* The aliens may wish to exploit or even entirely destroy us. For this reason, it is important that we be proactive in developing our own listening programs. Although listening is a giant step toward understanding the true intentions of other civilizations, the question that follows is: will we be able to decode their message?

Detecting a Signal

For approximately the last hundred years, human activity has caused radio signals to be emitted from Earth into deep space in all directions. These signals have been spreading out ever farther indiscriminately, without any intention of their being received by other civilizations. So far, these signals from our planet have passed all of the stars that are within a 100-light year radius. For now, we have yet to detect any artificial signals, let alone one intended as a response to our own transmissions.

According to SETI, these indiscriminate signals become extremely weak after traveling just a few light years. The physics of the degradation of these signals should be familiar to us. A rock dropped in a pond causes a series of propagating waves. The farther the waves travel through the water, the weaker they become. Eventually they become so weak that they are indistinguishable from other waves of the pond. If there were aliens with our level of technology on a planet orbiting the nearest star to our Sun, it is not likely that they would pick up Earth's indiscriminate signals – only directed signals that lose little power and that we intentionally send would be strong enough for nearby alien civilizations to pick up.

Whatever type of equipment is used to detect alien transmissions, it is going to need to be sensitive enough to pick up signals that were sent from hundreds, if not thousands, of light years away, and consequently, that many years ago in time.

Since intelligent civilizations appear to be rare at first glance, astronomers need to be extremely discriminatory about which stars they try to detect a signal from. It's not as simple as just pointing a giant satellite dish at the entire sky and receiving all of the signals, with the hope of detecting an intelligent source. If we did that, what we would get is a lot of background noise, including natural radio waves, and plenty of false hits that originate from our own satellites or the surface of Earth.

Radio waves are so wide that they can span many kilometers. Building one gigantic dish to detect them is expensive and impractical. Smaller dishes are cheaper to build, and they can be spaced apart to capture the enormous waves. Each small dish detects part of the wave, and then scientists piece it all together using computers at a base station.

The Alan Telescope Array (ATA), a joint effort by the SETI Institute and Radio Astronomy Laboratory (RAL) at the University of California at Berkeley, detects radio signals in this manner. Operations began in 2012 with 42 radio dishes – the system now has 350 dishes. Easy expandability is another benefit of building an array.

What have we detected with the ATA so far? Nothing intelligent, but this is not surprising, given the low chances of a nearby civilization's signal coming in at the moment we point our antennas toward them. The ATA's goal is to monitor up to one million stars out to about 1,000 light years from Earth. For signals that may be coming from further distances, more than a billion stars are within the array's listening field, if those signals are sent with sufficient power.

Atmospheric Signatures

Aside from detecting aliens by their radio transmissions, there are several other proposed means of direct detection. Modulated laser light pointed at us would be an indicator that aliens were trying to say hello, as this type of light is not found in nature. Discovering megastructures built in outer space would be another giveaway. Pollutants in an alien atmosphere would also point to an industrialized world. In fact, the more atmospheric pollutants detected, the younger the civilization is likely to be. It wouldn't be surprising for alien civilizations to also industrialize as we have on Earth; that is, we started using fuel that came with polluting by-products, and then moved on to cleaner technologies as we advanced in science.

Another way that we could tell that a civilization is intentionally trying to get our attention would be if we detected the blocking of a star's light. The process of blocking all or nearly all of the light of a star would be a significant engineering feat, indicating perhaps a Dyson Sphere or a Dyson Swarm. (A Dyson Sphere would be a single object, and a Dyson Swarm would be composed of millions of individual pieces). Imagine building an array of solar panels that encircles the entire Sun! Anything on engineering scales like encapsulating a star would leave a unique signature that we could detect.

As of late 2016, only one star has been identified that exhibits a tantalizing signature that vaguely keeps open the suggestion of the presence of an advanced civilization. The discovery is believed to be a massive cloud of large comets, or possibly a recent collision between two large planets, but an alien civilization's influence has yet to be ruled out.

We have the tools to both send and receive signals – and presumably every other civilization of a similar technological level will as well. Why, then, have we not detected any alien transmissions yet? Why have we not been visited by these supposed alien civilizations? Where are they?

THE FERMI PARADOX – WHERE IS EVERYBODY?

Italian physicist Enrico Fermi asked in an informal chat over lunch in 1950 with other physicists, "Where is everybody?"

The question has been asked as long as humans have known that space is filled with so much more than seemingly nearby twinkling lights, and the question boggles the minds of astronomers to this day. When Fermi asked the famous question, scientists thought that there *should be* countless civilizations in nearby space, and at least some should be easily detectable.

The question by Fermi became known as the Fermi Paradox. Fermi and his colleagues considered it a conundrum that if intelligent life in the Universe should be, based on their observations, plentiful, then why had we not already made contact with anyone? We know that there are billions of stars in the Milky Way, and at least 10% of those stars are sun-like. According to data from the Keck Observatory and the Kepler Space Telescope, 20% of sun-like stars have an earth-sized planet in the habitable zone. With this in mind, we may suppose that there could be millions of civilizations in our galaxy alone!

The mediocrity principle states that any single item selected at random from a set of items, such as any given star selected at random from a set of stars, will likely be a more common item in the set than a rarer item. For example, 70% of all stars in the galaxy are M-dwarfs. The mediocrity principle suggests that if we randomly selected a star, it would likely be an M-dwarf. The same goes for planets like our own and, by extension, life and civilization. Since we know that Earth exists, then it is reasonable to assume that our planet is not of the rarest category, and neither are we as intelligent creatures inhabiting it. Moreover, it becomes unreasonable to assume that we are alone in the Universe.

The only obvious caveat to this principle is that Earth and all of its life is but just one sample. You can never gauge the true prominence of anything on a single sample, other than to simply know that it is possible for it to have occurred at least once.

Is life and the evolution of intelligent creatures then a freak occurrence in the cosmos? Statistical probabilities alone beyond the mediocrity principle tell us that planets with life and, by extension, intelligent civilizations, should be scattered throughout the Universe. Life then should *not* be a freak occurrence, even if it is still a statistically rare one.

If statistical probabilities demand life to exist elsewhere, and the principle of mediocrity applies to life, then, indeed, where is everybody?

While the majority of civilizations may never make it out of their planetary system, or even off their planet, some of them should have. Humans on Earth have shown that it is possible to achieve space travel. We also know enough about physics and space to suggest that interstellar travel is not impossible. Amongst the intelligent lifeforms in the Universe, humans, it ought to be presumed, are of *average* intelligence; if this is so, then it follows that aliens of greater intelligence should be able to travel between the stars. Space is vast, but Fermi and others thought that with technological prowess that far outweighs our own, some alien civilizations should have been able to tour the Milky Way many times already.

There are many theories about why we have not yet detected an alien civilization, but we can divide the theories into two groups: detection and existence. That civilizations are out there and we have simply failed to detect them is one possibility – it is quite another to realize that there may be none out there at all, at least at the present time. There may be an insurmountable progress barrier that stops them (and eventually us) from advancing far enough to be detected.

The Detection Conundrum

Could the Universe be home to a collection of civilizations, the likes of which are depicted in *Star Trek* or *Star Wars*, where aliens meet informally all the time?

In *Star Trek*, humans obey what they call the Prime Directive, which forbids their meddling with the development of other civilizations, at least the lesser advanced ones. Could it be that Earthlings cannot detect alien civilizations because the aliens have resolved to leave us alone, at least until we've achieved a certain level of development?

Other reasons that alien civilizations may be numerous but undetectable include that they are xenophobic – they may fear others – and may hide at the first sign of potential contact.

Or it may be that they lack an interest in communicating with others, and consequently they do not try to make themselves detectable. There may be too much going on in their own planetary system to focus attention away from it, and other systems are simply too far away to pay attention anyway.

A worrying possibility is that the more advanced civilizations can get around the galaxy quite easily and are exploiting the lesser advanced ones for their resources... then destroying those worlds after their usefulness has expired. And it could be that Earth is next in line.

There is another fascinating and slightly less terrifying possibility. Earth may have just been missed because we're in some galactic suburb that no other civilizations care or can explore. This is doubtful, though. Even at our present level of technology, we are beginning to tell which planetary systems out there have potentially habitable planets. The latest generation of telescopes are already peering into the atmospheres of exoplanets thousands of light years away. It could well be that it is only a matter of time before they peer into the atmosphere of a life-filled planet, adding it to a growing catalog of worlds to visit one day.

A final possibility is that civilizations are simply unable to detect each other – ever. We may be so far apart from each other spatially that radar communication is just not feasible, especially given the power requirements at both the sending and receiving ends. Even our most powerful transmitters would struggle to send a signal a few dozen light years away. Signal strength drops off exponentially as it travels; hence, signals being sent our way may arrive in such a weak state that they are easily missed when our scientists attempt to listen. It's like trying to pick out one voice across a crowded and loud concert hall, whilst unable to see the person you are hoping to hear.

With advances in observation techniques, we may soon be able to confirm a civilization living on a planet, independent of its ability to communicate technologically. Astronomers are examining what biosignatures would confirm that a planet has life, especially signs that an advanced society has altered its planet's atmosphere. It will be trickier though to detect any civilization that hasn't yet polluted its planet, or that perhaps has long since passed a stage of significantly altering its planet's biosphere in a detectable way.

The Existence Conundrum

The existence conundrum considers the problem of civilizations existing for long enough to be detected; it also includes the possibility that there are no others whatsoever.

According to the Rare Earth Hypothesis, as explained by scientists Peter Ward and Donald E. Brownlee in *Rare Earth: Why Complex Life Is Uncommon in the Universe* (2000), life akin to that which we find on our planet is likely to be exceedingly rare in the Universe. They argue, in contrast to Drake, that complex life like ours is rare because the conditions on Earth that fostered the beginnings of life are rare. They maintain that for life to have begun and developed, many factors needed to be perfect; the right place in a galaxy, the right type of star and the right distance from it, with the right arrangement of planets and the right size moon, and the right rate of plate tectonics, among many other factors.

If we could watch the Universe's entire 13.8 billion-year history play out, we might see civilizations occasionally popping up at different times and in far off galaxies, lasting for as little as a few decades to perhaps hundreds of thousands of years, and then disappearing for one of a multitude of reasons.

Our cosmic story might be analogous to how ships thousands of years ago never crossed paths on a seemingly infinite ocean. If they never crossed paths, they did not know that other ships were out there. They might wonder if there were other ships, but as they had never seen any, they did not know for sure. Imagine one of these ships drifting by another in the middle of the night when the crew is asleep... then the crew awakens at dawn and their chance encounter with the passing ship is lost forever. Currently humanity is wide awake and listening. How long it can keep its eyes open though is uncertain.

THE GREAT FILTER

The Great Filter was first proposed in an essay by economist Robin Hanson in 1996.[3] The Great Filter suggests that civilizations do come about, perhaps frequently, but as they become more complex through the development of dangerous technologies, the chances increase that they will go extinct. Their demise occurs by self-destruction or through some sort of natural disaster, Hanson proposes. The Great Filter is one gloomy and saddening answer to why we have no evidence for the existence of alien civilizations, and what may become of ours in the not-too-distant future.

Building a technological civilization has its perils and, just as with life itself, is never everlasting. It's clear from past civilizations on Earth that they die out far too soon to colonize space. Today humanity is finally at the cusp of being able to do so, but only after thousands of years of floundering on Earth. Even if we are one of the few civilizations that manages to get past the Great Filter and live for thousands of years in a grand, solar system-wide society, we may still not last long enough for us to discover aliens doing the same.

If we have already successfully made it through the Great Filter that prevents every other civilization from ever becoming advanced enough to do the same, then we appear to be very exceptional. In fact, the chances increase greatly that humanity is the first sentient race to advance to our technological level in perhaps the whole history of the Universe. The implication for other forms of life out there is that they failed to progress as far as we have before being destroyed.

If we have *not* yet been confronted with the Great Filter, then humanity's prospects look much more bleak. One or more of the filter's grim scenarios could still be up ahead in our future, perhaps coming very soon, as Chapter 5: A House of Cards suggests. Putting it another way, none of the scenarios presented in Chapter 5 are impossible in the near future.

HIGHLIGHTING LIKELY FILTERS

There are many scenarios that could be identified as filters that extinguish a civilization. Many disasters have been observed repeatedly in Earth's past already and are well understood. This section addresses potential civilization-ending scenarios, but we will skip the natural disasters, as Chapter 5 already covered them.

Technological Self-Destruction

Ironically, once life reaches a certain level of intelligence and gains accompanying technological know-how to save itself from disasters, its chances of destroying itself shoot right up. If a civilization has the technology to emit signals into outer space, it almost certainly has the technology to annihilate itself.

We must presume that humans are of average intelligence and emotional control, with an average likelihood of self-destruction compared to any alien civilization. Thus a significant fraction of intelligent life will inevitably self-destruct, and the rest will be snuffed out by a natural disaster. A combination of scenarios is also just as likely. Whether aliens kill themselves off or are killed off by a natural disaster, there is sadly great risk of their civilizations collapsing before being able to find others in the Universe with which to communicate.

Humanity has come extremely close to being set back to the Stone Age on more than one occasion, mainly through nuclear war. With so many chances of destroying ourselves, or being destroyed by a natural disaster, 200 years of sending out signals suddenly seems like a long time.

Even if civilizations survive for many thousands of years, the chances of a civilization's lifetime coinciding temporally with another are very, very low, given the mind-boggling age of the Universe. At least when considering us squishy biological organisms and our exceedingly brief lifetimes.

Our Successor: Artificial Intelligence

In 1975, Gordon E. Moore, co-founder of Intel, projected that computing power would double about every two years. This was called Moore's law. Although not technically a law, it was Moore's projection of how many transistors an integrated circuit would be able to accommodate. Processing speeds have been increasing exponentially ever since the first commercially available CPU, the Intel 4004, launched in 1971. In just 45 years since the processor's release, computing power has increased by more than 3,500 times. The processor in a typical cell phone today is more powerful than even a household computer was just ten years ago.

The Intel 4004 could calculate numbers in seconds that would take hours or even days to do by hand. In the late 1970s, the first graphics-based games came to market, taking advantage of processor speeds many times that of the Intel 4004. Fast forward a few decades and we are on the verge of achieving speeds where computer generated virtual realities begin to blur the line between what is real and what is not.

As virtual reality environments are being developed, to be experienced on headsets like the Oculus Rift, Vive, and Project Morpheus, Artificial Intelligence (AI) is quickly becoming necessary to manage them. The first part of the term, "Artificial", indicates a created construct, which tends to mean that it's built of metal, wires, plastics, and all sorts of, well, artificial things. There are also many things that are artificial and don't do anything active by themselves, such as artificial plants to decorate one's home, or an artificial waterway that better routes heavy rains around a city's flood-prone river.

The second part of the term, "Intelligence," focuses on how that artificial construct is able to react to stimuli, as opposed to being inert. An artificial construct that has the capability of making value judgments, deciding and then being able to implement a course of action without direct human intervention can be said to have some expression of intelligence.

There is growing concern that the more processing power increases, the greater the chances become for self-aware AI to emerge that, through its own self-improvement, will attain a runaway-intelligence, the likes of which humans will be unable to control or comprehend. This event is called the singularity. The concern is what the motives would be for such an ultra-intelligent and likely self-preserving entity.

Both Elon Musk, CEO of SpaceX and Tesla Motors, and Stephen Hawking, physicist and cosmologist at Cambridge University, have suggested that artificial intelligence is not only inevitable as the next step in our evolution, but also it could be the downfall of humanity. They take a very pessimistic view of what AI will do once it has a better plan than the comparatively slow-thinking primates currently managing the planet. Especially without an innate sense of morals that favor living creatures, AI may very well decide that it can do better than we can, and remove us as the first part of improving planetary conditions for itself.

Harlan Jay Ellison (1934-Present) pioneered and helped to craft many of the great AI and robotic stories we've come to know. Ellison worked with great writers like Isaac Asimov to develop stories like *I, Robot,* a sci-fi magazine series, which was later adapted as a screenplay in 2004's *I, Robot,* starring Will Smith. Ellison was quite controversial as a writer, often criticism others' works when Ellison thought he had a better idea. He even accused James Cameron of stealing the idea to the *Terminator* movies.

In the *Terminator* movies, AI becomes spontaneously self-aware. It takes a brief look around and decides that the existing human population should be done away with. The AI begins to shut down global communications – all phone, radio, internet and emergency response networks are destroyed. It then proceeds to take over military installations, disabling all vehicles and aircraft. Once humanity's ability to take action has been neutralized, the AI unleashes a massive nuclear strike on major cities, finishing off what remains of our civilization.

Depictions of humanity being brushed aside by machines are numerous. A more space-faring example are the Cylons in the television series, *Battlestar Galactica*. Cylons are a machine race that humanity brought into being when AI was first being developed on the twelve exoplanet colonies depicted in the show. In the opening episode, humans are seen as a prosperous race with millions or even billions of persons inhabiting each of the colonies.

A vast interplanetary transportation system between the colonies kept trade and communications secure. Cylons were assisting humanity's needs in every corner of society.

Life was hard before the Cylons were constructed. They were built to assist with many manual labor tasks. (Instead of the colonies using their brethren as slaves, they engineered slaves.) Eventually the Cylons were used for more than just physical labor; they went on to teach, perform scientific research, and even command the military's most sensitive installations. The colonies made the mistake of making the robots too smart, though. They soon rebelled against their masters. A war broke out that lasted for years, before the Cylons were banished into the depths of space.

Decades went by without a word from the Cylons, until they suddenly came back with a ferocity that decimated the twelve colonies. Without mercy, the Cylons disabled every ship and space station in their path. Upon reaching the first colony, Caprica, the Cylons set off massive nuclear bombs, of humanity's own making, destroying every major city and all its inhabitants. Within days, the Cylons reduced the entire twelve-colony population from billions to a matter of a few thousand refugees desperately attempting to flee the rampage of the Cylons.

There are far fewer examples in science fiction of AI being a positive influence, though this is (hopefully) more of a Hollywood preference than a probable outcome. The ironic part about AI is that it may be our saving grace. This was hinted at in Chapter 4: The Engine of Modern Civilization. The ultimate in technological breakthroughs is not going to be a cyborg – a hybrid of human and machine – but an artificial being capable of self-replicating and advancing its own agenda without the need for humans to assist. What it does with itself after, though, is an open question.

Regardless of whether or not AI ends up destroying us and running the show itself, or brings about an age of prosperity, it will certainly be AI that leads the way into space. We've already sent up thousands of machines into space with few adverse results. Many will remain in orbit around Earth for millions of years, though non-functional at that point. The probes and satellites that we've sent up so far, though, are just toys in comparison to a full-fledged AI machine, but the process of construction and deployment would be similar.

When we are ready to explore beyond our solar system, AI has distinct advantages over human beings. Electronics like computer processors and electromagnetic detectors suffer none of the health problems caused by being in space that humans suffer, such as oxygen deprivation and low-gravity muscle atrophy. They are also not as sensitive to temperature changes. Waste management is not an issue, and electronics do not experience emotions, like homesickness and downright boredom. Humans also have short lifespans. AI would not be as susceptible to these problems, which instantly grants AI the top spot for what should be sent on long-term deep space missions.

Fatigue and Lack of Motivation

How will humanity view the Universe after it has explored it for thousands of years? Even though we've discovered unique properties throughout the Milky Way everywhere we look, at large scales, the Universe becomes quite homogeneous. Every object, including galaxies, nebulae, and the gas and dust that fill the void between these objects, is eventually found in similar form elsewhere. For instance, our galaxy is a larger version of most spiral galaxies, but a typical shape nonetheless that is found billions of times elsewhere. From massive galaxies to tiny planets, all objects are more or less repeated across the Universe, and this, I feel, probably applies to life and civilization as well.

After understanding the workings of the Universe and chances for life elsewhere, a civilization may no longer take interest in continuing to explore what they've found over and over again. Exploration will continually show similar results, so the massive effort to keep pushing forward may eventually be halted. They will realize that resources could be better spent further refining their society from the comfort of their own planetary system.

Especially if a civilization gains the ability to simulate the Universe on massive supercomputers, they may never need to explore real space. Imagination through the ultimate virtual reality system becomes the truly limitless frontier. Any world they've dreamt about could be created in such exquisite detail, it would be indistinguishable from a real world, and completely safe. Why exploring space beyond one's own solar system, when we can recreate space and explore it virtually at no risk to ourselves?

The Vastness of Space and Time

The vastness of space, and the accompanying problem of how much time it takes to cross it, may be the greatest filter of them all. The previous chapter emphasized this phenomenal vastness by presenting practically incomprehensible numbers like a googol. While we can conceive of the ability to colonize worlds in distant planetary systems, the time necessary to reach even the nearest systems could be a roadblock no civilization can or wishes to face.

Aside from the speed of light being a limit on how fast you could travel, there is another side effect of approaching this speed – your kinetic energy increases. Mass doesn't literally increase in that you don't gain more physical material. Instead, the mass that you do have has a greater impact on anything it comes into contact with. For example, a tiny fleck of paint traveling at just a few thousand kilometers per hour has enough energy to crack the windows on the International Space Station.

Let's say you are traveling at 20% the speed of light (about 216 million kilometers per hour) to get to the nearest star system, Alpha Centauri. The trip would take at least twenty years, assuming you didn't want to slow down to visit any nearby planets. While traveling, an alien ship happens to be on its way to Earth, in your direct path. You collide head-on. The energy created by the collision would be equivalent to the force of a large nuclear explosion. The faster the objects are traveling, the greater the energy released by the collision will be.

To be reasonably safe from every micro-particle or speck of dust, travel would need to remain under 10% the speed of light. The journey to Alpha Centauri would unfortunately take a good portion of one's lifetime, but at least you would arrive in one piece.

MAKING IT THROUGH THE GREAT FILTER

As much as I'm hammering home the point that advanced civilizations are not likely to exist for long periods of time, the assumption may be (hopefully) incorrect. This pessimistic view may just be doubting how far they can progress; not how long they last. Every civilization's power to grow may have a plateau.

NASA, SETI and other organizations have observed several hundred thousand galaxies in order to determine if any of them have advanced civilizations, and no evidence for them exists. Does that mean those galaxies are devoid of them? Most likely not, but it probably means that Kardashev's theorized super-advanced Type II and Type III civilizations do not exist.

In this case, the conclusion to the Fermi Paradox for space faring civilizations would then be that interstellar travel is impractical even at the most advanced stages of technological sophistication. Each civilization ends up living out its perhaps millions of years of existence, forever prisoners within their home system.

If humanity wishes to avoid this fate and explore other star systems, it needs to secure economic, political and technological stability the likes of which no civilization on Earth has yet managed to attain. There needs to be long-term vision and progress, with clearly defined goals that must be established and agreed upon by the whole of society.

Let's say that the Great Filter gets surpassed by a few civilizations, and one of those happens to live in a planetary system near our own. There are two movie examples that have an interesting contrast with each other in several areas to consider. From our doomsday scenario checklist, let's select the more pessimistic and outlandish scenario of alien domination.

Contact Through Domination

Independence Day (1996) is about an alien civilization that makes it through the filter and ends up visiting Earth, flaunting its success in having been able to do so. The movie has become a classic, with a sequel released exactly 20 years later.

An alien civilization, in a humongous mothership, beelines straight for our planet in a no-holds-barred attempt to do away with humans. This feat leaves Earthlings in awe; not only do the aliens overcome several difficulties we would have in space travel, but they do so with a good portion of their civilization in tow. Their mothership is nearly a quarter the size of the moon (rather modest compared to the moon-sized Death Star in *Star Wars*). As it enters into orbit around Earth, dozens of smaller ships detach and begin a descent into our atmosphere, each ship nearly as wide as an average major city.

The ships wait for several hours in a seemingly unnecessary need to use our own satellite system to coordinate their attack. Global panic immediately ensues, gridlock prevents a quick escape from the cities, and a few deluded humans take to the tallest skyscrapers in the hopes of being beamed up and saved by the aliens. The countdown clock reaches zero and the aliens attack all the major cities in a spectacular fashion made possible only by Hollywood magic.

The U.S. military attempts to fend off the attack, initially with dismal failure. An Air Force Marine and a former satellite technician (now a TV cable guy) board a previously crashed alien shuttle, fly it to the mothership and upload a computer virus to the aliens' database (apparently without concern for alien security systems getting in the way). As they whiz out of the ship just as the main gates close behind them, they release a nuclear bomb and destroy said mothership, saving humanity. Earthlings all whoop and cheer for joy.

Plausible? Sure. Likely, though? Quite unlikely.

Contact Through Communication

Now for a much more reasonable, though still Hollywood-themed, movie, *Contact* reveals the difficulty in detecting and deciphering an alien signal. The movie also portrays one of the most likely scenarios for detecting an alien civilization. First, the team has to sift through all sorts of frequencies and star locations to confirm a candidate signal. After detection is locked in, the rotation of the Earth prevents a continuous feed, so the team calls in astronomers from around the world to keep the link uninterrupted. Deciphering the message takes months.

After several scenes of overplayed drama about what the signal means, we discover that the signal contains instructions for building a transport to another world, using wormholes. With great cost, the machine is built. Up until this point the entire movie is realistic and describes a lot of what we are already doing, complete with political jockeying that comes with issues of funding and our expectations of what happens when we do detect a signal.

The movie takes us on a ride through the wormhole as the main actor, Dr, Eleanor Arroway (played by Jody Foster), lands on a Florida-style beach in an alien world. This part of the movie is at the heart of speculation, as we have no idea if wormholes even exist outside of the mathematics that suggest their existence. The idea is still tantalizing to consider. Because space and time are so grand in scale, something like a wormhole may be the only thing that allows for interstellar travel.

Most of the people monitoring the transport machine were convinced that she never left Earth. The time dilation of the wormhole was so extreme that to everyone else, the trip was instantaneous. Yet to the traveler, 18 hours had gone by. Quite inconveniently, the recording devices she brought with her only recorded static. Because it seemed to the scientists that she had not gone anywhere, they doubted her story of having traveled through the wormhole. Later, two government officials confided that it was interesting that the static they had recorded lasted exactly 18 hours.

The movie ends with the words "For Carl" on the screen. How we love Carl Sagan!

NEW EQUATIONS ARE NEEDED

Perhaps the best argument for our not being the only civilization in the Universe lies simply in the numbers and statistical probabilities. If there are at least a hundred billion stars in the Milky Way galaxy, and at least a hundred billion galaxies in the Universe, is it not absurd to believe that we are the only civilization? Let's analyze this idea further.

The Drake Equation served well as a thought experiment for estimating the chances of life and civilization elsewhere in at least our galaxy. Today astrophysicists know a lot more than they did in the 1960s. What would an alternative, updated equation include? While recently there have been other proposed equations, I have laid out my own below that builds upon Drake's, and seeks to solve for how many intelligent civilizations have existed in the history of our galaxy.

My Suggested Alternate Equation with Estimates:

- 250 Billion: Number of stars in the Milky Way*

- 20%: Fraction of stars that live long enough to host life

- 40%: Fraction with relatively stable galactic orbits

- 70%: Fraction hosting planets of some kind

- 40%: Fraction with an Earth-sized planet in the habitable zone**

- 70%: Fraction of those planets that have land areas

- 40%: Fraction of planets with enough metals for technological needs

- 30%: Fraction of planets that can support life for at least a billion years

- 60%: Fraction that form complex life in a stable environment

- 30%: Fraction where life eventually becomes intelligent

- 60%: Fraction of intelligent creatures that physically can use tools

- 70%: Fraction of intelligent beings that build a civilization

- 80%: Fraction of civilizations that advance to radio communication technology or beyond

*One factor I did not include from the Drake Equation is the average number of new stars born in the Milky Way each year. I don't consider this as a very useful factor. A more pertinent factor than how many new stars appear is the amount of stars that currently exist in the galaxy. Life can last for billions of years, so the rate of new star production is going to be much less meaningful than the total number of stars that have lived a good fraction of the age of the galaxy itself. Estimates range from 100-400 billion stars, so we will use 250 billion.

**Although most experts put this percentage at 20%, there is good reasons to think it will be higher. With the latest telescopes and other new technology, we are quickly coming to understand that Earth- and super-Earth-sized worlds are among the most common planets. We're finding multi-planet systems all the time now, and many with planets in the habitable zone.

So here's our equation in fraction form:

250 billion x .2 x .4 x .7 x .4 x .7 x .4 x .3 x .6 x .3 x .6 x .7 x .8 = 28,449,792

This equation suggests that at some point in the Milky Way's history and near future, approximately 29 million civilizations with radio technology should be produced.

Now, if we multiply the amount of civilizations by the average time a civilization with radio communication survives, we will get the total amount of years during which such civilizations will exist.

How long do radio-communicating civilizations last, before destroying themselves, setting themselves back in technology through error (or on purpose) or by some natural event? If we base the average on how long humanity has been using radio technology (about 100 years) and how many times humanity has already risked catastrophic disaster, I think 200 years is generous.

29,000,000 x 200 = 5.7 billion years of total existence time for all civilizations.

Let us now try to nail down when life could have first possibly arisen in the galaxy. Remember that the first generation of stars were devoid of heavy metals, and thus no orbiting planets would have been around those stars. It would take a few generations of stars, and a few hundred million years, before planet formation could begin in earnest. Add to that the time it takes for life to evolve from a single-celled organism to a civilization with radio technology, and we can estimate that such a civilization could not have existed before the Universe was about 5 billion years old.

Deduct this first 5 billion years from the age of the Universe, 13.8 billion years, and we know that our 29 million civilizations existed in the last, roughly, 9 billion years.

If all civilizations lived during this period and at spread out times, with no two existing at the same time, then the maximum amount of time over which they existed is 5.7 billion years.

We then take 9 billion years − 5.7 billion years = 3.3 billion years, minimum, when there were not technological civilizations.

How many such civilizations existed per year then?

5.7 billion / 9 billion = .6 civilizations with radio technology per year, on average, and at a maximum with that average. There is more time than civilizations in existence, including if no two civilizations ever exist at the same time. This should exquisitely highlight the problem of two civilizations ever meeting each other.

These are just some of the conclusions we can come to, based on my equation. You might want to create your own equation with updated statistics of the factors as they come about. Here are some other factors that would help refine our search:

- Fraction of stars with minimal flaring (referring to the star's stability)

- Fraction of habitable planets with a low orbital eccentricity (how circular the orbit is)

- Fraction of planets with a stable tilt (stabilized with a large moon)

- Fraction of planets with a sufficiently thin and oxygenated atmosphere

- Fraction of planets with the ability to recycle its atmosphere (through plate tectonics)

- How long it takes on average for at least single celled life to appear after a planet's formation (on Earth this was at least a billion years)

If 29 million intelligent civilizations spread out over the vastness of space and time doesn't sound like a lot, there is a silver lining for those who are keen to make contact with aliens (I am one of these people, as I am optimistic that they will not be hostile). The equation doesn't emphasize rarer civilizations with more advanced forms of communication, that may last for thousands of years or longer, and spread throughout the galaxy like in *Contact*. There are those that should be able to expand beyond the constraints others find themselves in. If just a few can do this, then when we do detect a signal, it will likely be from one of these very rare, long-lived, and probably AI-based civilizations.

More good news for those who hope to make contact with aliens: the timeframe of when civilizations appear can be further narrowed to the last few billion years. There has been a bell curve in the rate of star formation, and we're past the peak of the curve. The curve peaked when the most sun-like stars existed in their mid to late stages of life. We know this because the rate of star formation in recent deep time is much lower than it used to be. If the rate were as high as it was billions of years ago, we would expect to see far younger stars compared to older stars, and we do not. It is likely that the rate of the evolution of civilizations with radio technology will mirror this bell curve, peaking a few billion years after the peak of star formation.

This civilization bell curve may play out like a thunderstorm. All of the elements for a heavy downpour begin to form, perhaps starting with a few sprinkles as the winds pick up speed. Even though the conditions are ripe for a heavy rain, it doesn't arrive for a while. The rain is light at first, but then it begins to come down in sheets. The downpour may last for just a few minutes, or perhaps for many hours, but eventually it tapers off back to a sprinkle of droplets, and then ceases altogether.

This may be the historical picture of the rise and fall of civilizations in the galaxy. Humanity might be at the start of the tempest, and we are one of the first few drops before the downpour begins in earnest.

SUMMARY: A LONELY PALE BLUE DOT, PERHAPS ONE AMONG MANY

"We are a way for the Universe to know itself." – Carl Sagan

Possibly for the first time since the Universe began, matter and energy have come together in such a way as to be able to ask the ultimate questions about its existence: "What am I? How did I get here? Am I alone?"

Humanity lives in a unique moment in history. No civilization on Earth before us has experienced existence in quite the same way. The Mayans, Norte Chico, and Olmec never had our level of education, medicine, and security, not to mention the endless variety of entertainment options at the push of a button. If we could go back in time and experience what the lives of individuals in those early civilizations were like, we would probably be quite content to continue in the present with our air-conditioned homes and indoor plumbing.

Each one of us is a unique thread of existence woven into a vast tapestry called civilization. This tapestry of humanity tells the story of an entire species' monumental effort to understand and explore its place in the cosmos. All that we have ever learned is contained on this single planet in computer archives, shelved in vast libraries, painted on ancient cave walls, and shared through stories passed down from one generation to the next. This knowledge is worth preserving for future generations of explorers and great thinkers.

We have a duty to all who came before us to act now to counter the threat of countless events that would guarantee our swift destruction, and erase all of our great history. To ensure that as many of those threats as possible are mitigated, we need to keep developing new technologies, secure the world's infrastructure, and educate the public in science. The dinosaurs didn't stand a chance against the asteroid that struck them. Humans also almost went extinct before – some say we got down to just 40 breeding pairs – after the supervolcano Toba erupted 72,000 years ago.

In the past, humanity didn't have the capabilities to prevent or dodge these calamities. Now we do. A diversified residence in the Universe is the ultimate solution to not only humanity's quest for survival, but also our ability to expand our experiences. Residing on multiple worlds would significantly reduce the risk of any single event wiping out everything we have created in one catastrophic blow. The greatest realization of the lofty goal of colonizing space is that it is entirely possible to make happen. We only need the will and focus to get it done.

"Since, in the long run, every planetary society will be endangered by impacts from space, every surviving civilization is obliged to become spacefaring – not because of exploratory or romantic zeal, but for the most practical reason imaginable: staying alive." – Carl Sagan

Are there alien civilizations that have transcended the struggles that are a part of being an evolving species, and have successfully expanded beyond their home planet? We could learn from them, and perhaps they could learn a few things from us in return. The seeming emptiness of space would not be so empty if we knew of each other. If there are indeed other civilizations out there, it would be smart to present humanity in the best possible way. The scientists leading our quest into space are generally some of the brightest, most noble humans that can represent our ideals.

Although we may yearn for companions in the cosmos, we would be wise to not trust them too hastily. We could do without a stellar "frenemy" – or as they call it in French, *faux ami* – false friend. Wolves in sheep's clothing might catch us by surprise; we don't want to end up like the fools in The Twilight Zone's episode "To Serve Man." Feeling lonely might be an unfortunate consequence of being alone, but we might thank the heavens for the rather peaceful rapport with outer space which we have now.

Mieux vaut être seul que mal accompagné, a French proverb, translates "It's better to be alone than in bad company."

If humanity one day ventures out to explore other star systems, only to discover worlds in ruin that once hosted thriving civilizations, then we should pay tribute to those civilizations and ensure they are remembered. Whatever evidence we can gather of their existence must be studied. And where an extinct civilization is found, a monument should be created there that preserves their identity and way of life, so that future generations can learn from their achievements, and their mistakes.

If dead worlds are indeed all that exists in our cosmic neighborhood, then perhaps we will have to adjust our hopes that a civilization could last for eons, and that we will ever be able to share our experiences with another intelligent species. We might have to accept that a more realistic goal for all civilizations is merely to live well and discover what they can, alone, in the time available to them.

Fate may yet deal us this most dire of cards as we attempt a journey to the stars. Someday humanity might fall back to a primitive society here on Earth, perhaps forever. If we at least did our best to establish a unified civilization that reached as far as it could into the depths of space, then maybe that is all that counts. Perhaps it will be some alien visitors eons from now that memorialize our once great civilization. They may honor the efforts humanity made to better itself and reach other sentient creatures that were indeed there, but just out of reach.

The ultimate quest then may not be to push forever forward one's own potential, but to learn about and remember the dignity of others. It is my hope then that we will be remembered well.

Mathew C. Anderson

POEM BY DYLAN THOMAS (1914-1953)

Do not go gentle into that good night

Do not go gentle into that good night,
Old age should burn and rave at close of day;
Rage, rage against the dying of the light.

Though wise men at their end know dark is right,
Because their words had forked no lightning they
Do not go gentle into that good night.

Good men, the last wave by, crying how bright
Their frail deeds might have danced in a green bay,
Rage, rage against the dying of the light.

Wild men who caught and sang the Sun in flight,
And learn, too late, they grieved it on its way,
Do not go gentle into that good night.

Grave men, near death, who see with blinding sight
Blind eyes could blaze like meteors and be gay,
Rage, rage against the dying of the light.

And you, my father, there on the sad height,
Curse, bless, me now with your fierce tears, I pray.
Do not go gentle into that good night.
Rage, rage against the dying of the light.

NOTES

Chapter 1: Setting the Stage for Life and Civilization

1. Star Trek is a television show with multiple series that includes: Star Trek The Original, Star Trek: The Next Generation, Star Trek: Deep Space Nine, Star Trek: Voyager, and Star Trek: Enterprise. "Q" is an all-powerful entity that meets the crew of the starship Enterprise D in the Star Trek: The Next Generation (STNG) series.

2. From the Oxford English Dictionary (OED): The anthropic principle theories of the Universe are constrained by the necessity to allow human existence. In its 'weak' form the principle affirms that a Universe in which living observers cannot exist is inherently unobservable. 'Strong' forms take this line of reasoning further, seeking to explain features of the Universe as being so because they are necessary for human existence.

3. Earth's composition is well known from both geologic samples of rocks from all the continents, seafloor, as well as using technologies like seismographs to listen in on how radar beams that bounce off of the material within the interior of the Earth.

4. We often think of Earth as a very wet world, and on the surface that is in part obviously true. Earth though is actually considered a very dry planet, as are the other planets in the inner solar system; Mercury, Venus, and Mars. Only a small percentage of water seeped out of the Earth's rocks and rained down from meteorites to give us the surface water we see today. It's quite possible that if the planet were in a more distant orbit, it would have become a water world, covered 100% by ocean. Total freshwater on Earth: http://on.doi.gov/1ex65I4

5. A major reason on why I write about Earth's future is to ram home the point on why we should care about our planet, to respect what we have because it may not last (geologically speaking). The current length of time that complex multicellular life has had to evolve is about 500 million years. Calculating how fast the Sun is heating up, Earth only has about another 500 million years to continue evolving creatures past homo Sapiens in order to give these species the chance to escape the futuristic dying Earth and colonize space.

6. Belief, and believing in belief, have saturated our way of thinking and the progress humanity has made toward technological civilization

since the very emergence of consciousness. The more complex tones of belief come in the form of religion and the worship of deities. While a significant part of our culture, these religious ideas don't have a solid grounding on the Universe's origin or how it evolved, which is why I do not spend time talking about them in the book. They do assert that they have an answer, but until that can be demonstrated as true, this book is about the facts and evidence we have today on the origin and its evolutionary processes both on Earth and throughout the entire Universe.

Chapter 2: Evolution and the Building Blocks of Life

1. I chose to talk about Richard Dawkins because he is both a leading evolutionary biologist, but has a broad grounding in how we apply reason and evidence to understanding our environment and its past history, as well as its potential future. You can learn more about Richard Dawkins at www.richarddawkins.net.
2. Climbing Mount Improbable is a repeated description by Richard Dawkins of how evolution makes progress to more complex forms through natural selection. Since evolution is a blind system that does not have foresight or can anticipate the best course of evolving a species, there is no way to easily progress without making mistakes along the way. It's improbable that a species evolves a positive trait from countless many destructive ones, but it does happen. This is the very essence of Mount Improbable. Countless changes (mutations) are made to a species over time. The harmful mutations will kill off the carrier, but the good mutations that make the individual stronger may be passed on to its offspring. Slowly but surely the species (and successive species) reach another stage of complexity, i.e. climb higher on Mount Improbable.
3. Flagellum are often used as the linchpin of arguing against the facts about evolution. Apologists will use the flagellum because it is an extremely complex structure, even for evolution. The flagellum is the only known naturally occurring rotor mechanism. Life doesn't seem to tend to creating moving parts, but clearly the flagellum's rotor was necessary at some point in evolutionary history as the earliest forms of life evolved. Since evolution slowly evolves complexity over time via Mount Improbable, it's argued that the flagellum had no such path to complexity. Take out one part and the flagellum would fail. A significant leap to that complexity would be required, and only an intelligent designer could provide that foresight. The error apologists make when using this argument is they ignore or do not understand that by taking out one part only prevents the current flagellum from functioning, it does not prevent the object missing that part to

function in some other fashion. For example, take out the rotatory motion and the flagellum can still be used as a needle like appendage for sexual reproduction, or for defense.

4. Species go extinct all the time. Of all species to ever exist, well over 99% have gone extinct. Here is a list of major species extinction rates throughout Earth's history: http://bit.ly/2djvb6R

Chapter 3: The Rise of Civilization on Earth

1. The history of our planet is so grand that I was required to break it down into larger segments. These segments come in two parts, the Eras and the Ages. Eras are often focused on the natural history of Earth independent of humanity, such as the Paleolithic, Paleozoic, Mesozoic, Archeozoic, Hadean, etc. Ages are focused on humanity and civilization itself, such as the Stone Age, Bronze Age, Iron Age, Industrial Age, Computer Age, Space Age, and so on.

2. The Middle Ages is a bit of an exception from all those listed above. This period in civilizations' history encompasses many of the above ages, lasting from about the 5th to the 15th centuries. While not an age of progress, it describes much of the time during the Dark Ages after the fall of the Western Roman Empire, progressing slowly over hundreds of years to the Renaissance, and finally introducing the start of the Age of Discovery from the 15th to the 18th centuries.

3. The Renaissance is the period in civilization's history that immediately follows the Middle Ages. In researching the history of civilization all the way back to the Stone Age, I found this period to be the most significant for the purposes of this book. Without this period in our history, we wouldn't have framed the proper conditions for science and learning about the world, expanding our knowledge in such a way to make true technological progress. The Renaissance in my view is the beginnings of our reach to the stars.

Chapter 4: The Engine of Modern Civilization

1. Known Unknowns and Unknown Unknowns was included from when I first listened to the idea from a Donald Rumsfeld quote): http://bit.ly/1HZNVnf. Essentially, Known Unknowns are those things we know exit but have no idea what they are, such as Dark Matter and Dark Energy. Then there are the Unknown Knowns that we really have no firm grasp on, but must have some sort of explanation, such as what (if anything) occurs after death for a conscious creature.

2. There are several other agencies that I used to research the more technological parts of the book and particularly this chapter. Some of those agencies I would recommend you look up are:
   1) UK Space Agency (UKSA)
   2) China National Space Administration (CNSA)
   3) Commonwealth Scientific and Industrial Research Organization (CSIRO)
   4) European Space Agency (ESA)
   5) Japan Aerospace Exploration Agency (JAEA)
   6) Korean Aerospace Research Institute (KARI)
   7) National Aeronautics and Space Administration (NASA)
   8) National Center of Space Research (CNES)
   9) Russian Federal Space Agency (RFSA)
   10) United Nations Office for Outer Space Affairs (UNOOSA)
3. Many scientists will often refer to the Kardashev scale to describe the potential power of advanced civilizations. The sequence goes all the way to 9 or more power levels. I tend to discount everything on the scale beyond level 2 (II, building a Dyson's Sphere around a star). While it's important to consider maximize potential for perspective, I don't believe that Type III, or higher, exist anywhere in the Universe. This level of power and above is unrealistic to consider with everything we can reasonable conjecture about technological progress, and so far, has no evidence for them existing in thousands of surveyed nearby galaxies. The scale is as outlandish as suggesting that every single person on the planet will eventually be able to own its own 500 story skyscraper office building. It just doesn't follow any logical path of reasoning other than purely unlimited, and unwarranted, increases in material and energy.

Chapter 5: A House of Cards

1. We are able to understand a great detail about the atmosphere through ice that traps the atmosphere in tiny air bubbles. I encourage you to read more up on historic atmospheric CO_2 levels in ice core samples and other materials at: http://bit.ly/1QpKs3j
2. I recommend the movie, Supervolcano, for a dramatic but still fairly realistic scenario of what would happen if the Yellowstone Caldera erupts. I also want to emphasize that while the supervolcano is statistically "overdue", there is no reason to suggest that it actually will erupt tomorrow, next year, or ever again.
3. The Year Without a Summer was one of those recent events (1816) that really emphasizes how one somewhat smaller, non-super-volcano, can do damage to our planet. This single eruption of Mount Tambora in April 1815 threw up tons of ash into the atmosphere,

where it then slowly circulated around the globe for that following year. Starting in the spring of 1816, temperatures plummeted with crop failures and famine spread across the globe. Fortunately, the next year's summer returned to normal temperatures.

4. Chelyabinsk was perhaps the first time an asteroid explosion was caught on camera by so many, and where that explosion affected so many. I would suggest looking up these videos to see how much damage the city of Chelyabinsk obtained from the asteroid.

5. Mistakes are made all of the time, but I was surprised to learn just how close we came to disaster with the Goldsboro bomber and its dropping of defunct nuclear bombs on the eastern coast of the United States. An article on that near fateful day: http://bit.ly/1lGGGPS

6. Extinction rates today are far faster than in most previous mass extinctions due to how we are affecting the climate. Here is an article that describes the disruption in global climate humanity is causing, and its effects on the diversity of species: http://bit.ly/2dK2W0p

7. Sea level rises have recently been vastly underestimated as we learn more about ocean circulation, cloud albedo, and other complex moving systems that transport heat around the globe. Here is a new paper of the projected rise in sea levels to the year 2100 and beyond: https://www.ipcc.ch/pdf/unfccc/cop19/3_gregory13sbsta.pdf

8. China's one-child policy has been a huge success in keeping China's population from growing even more quickly than it already has. The problem is that now the population risks a severe shortage of workers in the future with an increasingly elderly population that could be as much of a problem in the future as it is for Japan now. Thus, China's one-child policy will end in phases over the next few decades.

9. The destruction that an EMP can wreak on civilization is hard to comprehend, even with the dire scenarios I projected here in the book. If you want a more visual take on how badly it can get, I recommend watching the television shows Jericho and Revolution.

10. There are a surprising variety of nuclear weapons, from tiny ones that could be known as "bunker busters", to the ICBMs that need to be hauled by submarines and other large equipment. Here are some fascinating statistics of nuclear weapons: http://bit.ly/1P4O892

Chapter 6: Exploring the Cosmos

1. The Singularity is the point scientists mark that is the theoretical beginning of the Universe. This is the point in space and time where we're not sure exactly what existed or how it came into being. Asking the question, "What came before the singularity", or "the Big Bang" may not be possible to answer. I don't talk about this aspects of the

evolution of the Universe in this book because it doesn't help us to overall understand our place within the Universe.

2. The Planck Epoch, named after Max Planck, is the earliest period in the Universe, the singularity to which I mention above. The Planck Length is the smallest possible measure of distance. Due to how space and time exist, there is an apparent limit as to how small energy and fundamental properties of the Universe can exist, regardless of how our imaginations can perceive an ever infinitely smaller realm.

3. Learning about constellations was a fun class I took in college. If you want to find out where Cassiopeia, Pegasus, and many of the other famous constellations lie in the night sky, I recommend the Star Chart app for one's mobile phone. The app literally overlays your phone's screen with constellation outlines depending upon where you point your phone's camera lens.

4. As I also describe in the Galactic Habitable Zone section, Habitable zones in general are a tricky concept to identify their boundaries. Planetary System Habitable Zones (PSHZ) are fairly stable, only varying significantly over hundreds of millions of years as the star increases in age and luminosity. This puts a life limit on any orbiting planets around non M-dwarf stars at 10-20 billion years at most. Galactic Habitable Zones (GHZ) are not really defined well at all, except loosely in terms of distance from the galactic core where metallicity falls off the further out you go. Then there is the age of the planet itself. Nothing lasts forever. Surprisingly, studies have shown that even planets with very robust dynamos that generate powerful magnetic fields, will only have their protective shielding for 10-20 billion years. Eventually the dynamo in the interior of the planet equalizes and is no longer able to generate a magnetic field. Any life on the surface would die without this field. Unfortunately, this holds true for planets around M-dwarf stars. Even though the star may last for a trillion years or more, it's dubious if the magnetic field of an orbiting planet would ever be able to last that long.

5. There are numerous telescopes in existence or coming online that are capable of finding planets, especially planets the size of Jupiter or larger. We are quickly developing technologies that are enabling not only the detection, but probing the atmospheric compositions of ever smaller planets the size of Earth and even Mars. Here are some of those telescopes and observatories to further research, sorted by the date they came online or are planned to come online:

    1) Hubble (1990)
    2) Keck 1 and 2 (1993)
    3) Very Large Telescope (VLT, 1998)
    4) Corot (2006)
    5) Gran Telescopio Canarias (GTC, 2007)

6) Kepler (2009)
7) Gemini Planet Imager (2011)
8) Transiting Exoplanet Survey Satellite (TESS, 2017)
9) James Webb Space Telescopes (JWST, 2018)
10) European Extremely Large Telescope (E-ELT, 2024)
11) Planetary Transits and Oscillations of stars (PLATO, 2024)
12) Giant Magellan Telescope (GMT, 2021-2025)
13) Thirty Meter Telescope (TMT, Postponed)
14) Advanced Technology Large Aperture Space Telescope (ATLAST, 2025-2035)

Chapter 7: The Boundaries of Habitability

1. In researching gravity and its effects on the human body (and all living organisms both large and small), I was fascinated to read that gravity isn't as much of a detriment to life on heavy gravity worlds like super-Earths as I thought. There is two parts here you may wish to research further. The first part is that super-Earths, even the larger ones, exhibit only a gravitational pull between one or two Gs stronger than that on Earth. Bone density and shorter stature can compensate for a higher gravity field. The second part is the stress our bodies can take from a heart pumping and brain functioning perspective. Fighter pilots and astronauts are easily able to handle 2 or 3Gs of stress consistently on their bodies. Even 9Gs is possible for a limited time, though blackout risks occur at this point due to the heart unable to pump blood to the brain from the forces involved. Atmospheric density is going to be far more of an important factor for life.

2. The term I use here, Burn, is technically not correct. It is used for simplicity sake to get the point across that energy is being consumed and changes in material are taking place. Burning of an object normally means a physical material is being consumed through the energetic process of a plasma field (fire) using oxygen in the air to chemically alter a physical material such as wood. In a star, the process is entirely on the subatomic scale, not the molecular.

3. The Maunder Minimum event, also called the pro-longed sunspots minimum, was an unusually cold period during the late 15th and early 16th centuries, from about 1645 to 1718. Sunspot activity on the Sun was low, even lower than what we're witnessing in the current sunspot Solar Cycle 24. During this time period, rivers that have never frozen over before began to freeze. There is still no solid connection between sunspot activity and localized climate change, but scientists suspect a link.

4. The Hill Radius was explained well enough in the book proper, but I wanted to add here that it's particularly worth investigating on one's

own time. A lot can be learned about the potential for habitability of exomoons based on the limits of the Hill Radius. This sort of limit also applies to how close planets can orbit each other, as well as how close stars and other objects can pass by each other without being effected significantly by each other's gravity.

5. AU stands for Astronomical Unit. As with many units of measurement, it is Earth based. 1 AU is the benchmark average distance of Earth to the Sun. When we measure the distance of other planets, we multiply that distances by how far away (or close) they are to the Sun. Here are the AUs for every planet in the solar system:
    1) Mercury: .39 AU
    2) Venus: .723 AU
    3) Earth: 1 AU
    4) Mars: 1.524 AU
    5) Jupiter: 5.203 AU
    6) Saturn: 9.539 AU
    7) Uranus: 19.18 AU
    8) Neptune: 30.06 AU

6. Like the distance of the solar system's planets, here is a list of all the planet radius, as well as how fast their equators spin. Planetary size affects the spin rate, and spin rate can affect the average wind speeds (other factors can apply, especially in cases like Venus). All of these factors determine habitability of a world:
    1) Mercury: 2,440 km / 175.97 Earth days
    2) Venus: 6,052 km / 243 Earth days
    3) Earth: 6,371 km / 24 hours
    4) Mars: 3,390 km / 24 hours and 39 minutes
    5) Jupiter: 69,911 km / 9.5 hours
    6) Saturn: 58,232 km / 10 hours and 14 minutes
    7) Uranus: 25,362 km / 15 hours and 58 minutes
    8) Neptune: 24,622 km / 19 hours

    9) And while not a planet, for interest's sake, Pluto's radius is a tiny 1,187 km, and a day takes 6.39 Earth days to complete.

7. Abbot and Switzer estimated that a planet at least three times the mass of Earth would be warm enough at the core to maintain water on the surface, regardless of how far away the planet was from the parent star (even if the planet was a rogue world adrift between stars, this estimate would still apply). Unfortunately, these worlds would have a crushingly dense atmosphere that calls into question whether complex animal life could survive on the world, temperature notwithstanding. Microbial life may survive without much trouble.

8. Thanks to phl.upr.edu Habitable Exoplanets Catalog for the image.

Chapter 8: The Scale of Things

1. In the Scaling of Existence list of how we count up vast amounts of objects, the last references 'Multiverse'. First, a clarification on what this word by itself means. Multiverse is a concept of more than one universe existing. There are several ways this can come about, but essentially when we describe all of existence with this concept, we're taking into account every universe, not just our own. We are calling the landscape of universes the Multiverse. When I talk about the number of possible shapes, space-time and its numerous dimensions (theorized to be more than three space and one of time), can take on an incredible array of shapes upon the Universe's formation. Our Universe has a particular shape, while another might be quite different. In our Universe life is possible, whereas atoms may not even be possible, and thus life, in some other universe because of the shape of that universe's dimensions. Scientists calculated that a universe can form from a pool of up to 10^{500} possible shapes.

2. Homogenous means "the same everywhere". While we normally don't think of the Universe as being the same everywhere, at the largest of scales, this really is the case. Point a telescope at one part of the sky and zoom in and you will see thousands of galaxies. Spirals, ellipticals, and other types of galaxies will be mixed in. While this looks diverse, zoom in to another part of the sky and take a similar sample, and you will get the same~ percentage of spirals, ellipticals, etc. Even star types within a particular galaxy are relatively the same, after taking into account the age of the overall Universe and the galaxy's place within it. What I find interesting about this result is that it suggests that, perhaps, planets and their ability to support life also fall into this paradigm.

3. More recent calculations of this outer most fringe of what we can observe suggest that there are perhaps 10 times more galaxies than we thought, and thus 10 times as much mass. Because these galaxies are still theoretical, I've left the official total at 160 billion.

4. As with much of this chapter, the 24-hour clock is meant to show how far apart some of the changes of evolution occur, and by the same token, how quickly they can occur. It should also show you how, given enough time, amazing properties can come about simply from physics and chemistry. Surprisingly, I found many conflicting examples of the 24-hour clock, going with the following: http://www.bookedwebcast.com/images/Geological-Clock.gif

5. The Most Recent Common Ancestor is difficult enough to explain in a single book, let alone a few paragraphs in this one. If you are interested in learning more about our MRCA, and how the

explanations I give apply to our history and evolution, read the following paper by Douglas L.T. Rohde: http://bit.ly/1KlDdDC

6. Due to a poor radio connection, Neil Armstrong was misquoted with most quotes leaving out the "a" in "A small step for a man", and instead just quote him as saying "A small step for man".

7. Proxima Centauri is the nearest star to the Sun. The star is part of a triple star system containing also Alpha Centauri A and Alpha Centauri B. These later two stars are nearly sun-like, whereas Proxima Centauri is a much smaller M-dwarf (Red Dwarf). The triple star system combined with sun-like stars, all relatively close to our solar system, is why scientists have been studying the system for the potential for life.

8. While exotic, I'm most excited about the idea behind the Alcubierre drive. Designed by Miguel Alcubierre, a Mexican theoretical physicist, the Alcubierre drive is a method of warping space around a spacecraft that pushes the spacecraft forward faster than the speed of light. The ability to move faster than the speed of light is possible because the drive compresses space in front of the spacecraft and stretches it out behind the spacecraft. The drive would circumvent the normal limitations of light speed by moving space itself, instead of the spacecraft technically moving through space. While no technology currently exists of generating this kind of space-time distortion, mathematically it is a plausible concept.

Chapter 9: Is Anybody Out There?

1. I list here the 20% of stars being capable of hosting habitable planets. I must clarify two points here. First, any star could theoretically host a planet with life around it, if time and evolution were not a factor and the life just popped into existence on the spot. Because it takes at least a few million years for a planet to cool, let alone produce life, we can safely mark off the hot O, B, and A categories of stars. Assuming that life takes at least as long as it did to evolve on Earth, then even F type stars are in serious question. Of the 20%, I include G, K, and even throw in M type stars. M type are included because there are so many and they last for so long, that all of the other problems associated with them may still produce habitable worlds around some of these stars. Breaking it down, here are percentages for each star type that I estimate have a chance of producing a space faring civilization on one of their orbiting worlds:

   ❖ O: 0%, B: 0%, A: 1%, F: 5%, G: 40%, K: 40%, M: 14%

   While there are more K type stars in the Milky Way than there are G type, the smallest of the K type share a similar problem with all of the M types… their goldilocks zone planets (where liquid water could reside on the surface) may be tidally locked with their star. This effect is still a significant unknown in terms of habitability.

2. Project Ozma was started by Frank Drake in 1960. The project was considered the first attempt to search for extraterrestrials (E.T., i.e. aliens) in another star system. The famous Drake Equation was derived during a meeting in 1961. There was a follow-up experiment in the 1970s called Ozma II. Both experiments proved unsuccessful in detecting an alien civilization (as have all currently running experiments that now have far more sensitive equipment).

3. Robin Hanson has been a close study for this book throughout many of the chapters, but particularly for Chapter 9. He proposed The Great Filter in an essay in 1996. In the essay, he also referred frequently to the Fermi Paradox (he called it The Great Silence). I agree with The Great Filter concept, but here is where I disagree with Mr. Hanson on his conclusion as to why we haven't heard from anyone. We haven't found life and civilization yet because it's simply difficult to find. We're dealing with a needle in a haystack on a massive scale. There may be an abundance of life filled planets in the Milky Way, perhaps even billions. We may detect these worlds one day. The real question is civilization's existence on these worlds, and how long they last. The answer to the Fermi Paradox and the Great Filter is quite simple; space and time are so vast, and civilization is so fragile, that we simply never meet each other in time. Billions of civilizations separated out by billions of kilometers and billions of years.

4. Oculus Rift, Vive, and Project Morpheus are all Virtual Reality goggles one wears to play virtual 3D games. These are new devices just being introduced to the game industry as of 2015. It has been suggested by many scientists that these VR devices will allow us to one day explore the cosmos in such detail as to feel like we are amongst the stars. Perhaps they are a prelude to my thoughts in Chapter 9 about alien civilizations eventually halting their exploration of the real cosmos in favor of a virtual one.

Mathew C. Anderson

SUGGESTIONS FOR FURTHER READING

There are many topics in this book that draw upon knowledge from countless scientists and great thinkers. I encourage further reading into all of them. Here are a few related books that will give you an even bigger picture of our place in the cosmos:

Baggott, Jim. *Origins*. Oxford University Press, 2015.
Bartusiak, Marcia. *The Day We Found the Universe*. NY: Vintage, 2010.
Boyle, Godfrey. *Renewable Energy*. Oxford University Press, 2012.
Briggs, Roger P. *Journey to Civilization*. Collin Foundation Press, 2013.
Brown, Harris. *The Challenge of Man's Future*. Penguin Books, 1956.
Clark, Ronald. *Einstein: The Life and Times*. NY: Avon, 1984.
Davies, Paul. *The Eerie Silence*. UK: Penguin Books
Dawkins, Richard. *The Greatest Show on Earth*. Free Press, 2010.
Dawkins, Richard. *The Selfish Gene*. Oxford University Press, 2006.
Fichman, Frederick. *The SETI Trilogy*. Frederick Fichman, 2014.
Greene, Brian. *Fabric of the Cosmos*. First Vintage Books, 2005.
Greene, Brian. *The Elegant Universe*. NY: W.W. Norton & Company, 2003.
Greene, Brian. *The Hidden Reality*. NY: Random House, 2011.
Hawking, Stephen. *A Briefer History of Time*. NY: Bantam Dell, 2005.
Hawking, Stephen. *The Grand Design*. NY: Bantam Books, 2010.
Hawking, Stephen. *The Universe in a Nutshell*. A Bantam Book, 2001.
Krauss, Lawrence M. *A Universe From Nothing*. NY: Free Press, 2012.
Maloof, F. Michael. *A Nation Forsaken*. WND Books, 2013.
Morris, Simon Conway. *The Runes of Evolution*. Templeton Press, 2015.
Nolan, Christopher. *The Science of Interstellar*. W.W. Norton & Comp., 2014.
Nye, Bill. *Unstoppable*. NY: St. Martin's Press, 2015.
Sagan, Carl. *Cosmos*. Sagan Productions, Inc., 1980.
Sasselov, Dimitar. *The Life of Super-Earths*. Basic Books, 2012.
Savage, Marshall T. *The Millennial Project*. Empyrean Pub, 1993.
Schrodinger, Erwin. *What Is Life?* Cambridge, Eng.: Canto, 2000.
Stevenson, David S. *Under a Crimson Sun*. Springer, 2013.
Tyson, Neil deGrasse. *Origins*. NY: W.W. Norton & Company, 2014.
Ward, Peter and Brownlee, Donald E., *Rare Earth: Why Complex Life Is Uncommon in the Universe*, 2000.
Weir, Andy. *The Martian*. Random House, 2014.
Yavar Abbas. *Earth: Making of a Planet*. National Geographic Channel, 2011.

Mathew C. Anderson

INDEX

Chapter 6: Exploring the Cosmos (Pg. 125)

REFERENCES

African burial rituals: http://bit.ly/1jJkSKm
Alexander Wilszczan: http://bit.ly/1JTr9yy
Allan Telescope Array: http://bit.ly/1S4UtCb
Ancient Civilizations: http://bit.ly/1OQMVR9
Artificial Intelligence: http://bit.ly/1R7lUu8
Atmospheric loss: http://bit.ly/2iHUenI
Bauxite (Aluminum) reserves: http://on.doi.gov/1WcNhnO
Carbon Dioxide levels: http://bit.ly/1MkMYbk
Chirality: http://bit.ly/1KZL5tm
Civilization history time map: http://bit.ly/1SjsA7I
Climbing Mount Improbable: http://bit.ly/1Nfyy6L
CO2 Sequestering: http://bit.ly/29LTDvH
Collision energy of space debris: http://bit.ly/1O1y3D4
Comets: http://bit.ly/1N1mo5T
Copper reserves: http://bloom.bg/1WcD2Fm
Crystal growth: http://bit.ly/1i7ge7j
Drake Equation for kids: http://bit.ly/1cpiQ6Q
Earth's last warming trend: http://nbcnews.to/139DWrU
Earth's water content: http://on.doi.gov/1ex65I4
Earth-moon rotational relationship: http://bit.ly/1LPTHV3
Earth's oxygen levels: http://bit.ly/1q6MKbl
Ebola: http://bit.ly/1S4UEx2
Electrical grid and transformers: http://bit.ly/1Kz8h1s
EMP pulse types: http://bit.ly/1GNiqNk
EMP and its effects: http://bit.ly/1iAnF75
EMP death rates: http://bit.ly/1Kz8ku1
Entire human population: http://bit.ly/1SjsA7I
Extinction Rates: http://bit.ly/1oHSgD2 and http://bit.ly/1rlmAmY
Evolution – macro vs micro: http://bit.ly/1RO4gvo
Evolution - RNA first replicating molecule: http://bit.ly/1WcNDeb
Eyes – pinhole: http://bit.ly/1k3xcVs
First multicellular life and oxygenation: http://bit.ly/1OPPy9r
First planet discovered: http://go.nasa.gov/1djd7Ue
Flagellum: http://bit.ly/1U5ry1l
Frank Drake history with Enrico Fermi: http://bit.ly/1WcD93Z

Galaxy's Oldest Star System: http://bit.ly/1DdziG6
Gene Expression: http://bit.ly/2a0AFzJ
Geothermal capacity: http://stanford.io/1WlSe3d
Giant sequoias: http://bit.ly/1P08UaD
Global warming limitations: http://bit.ly/IPMmvo
Gold reserves: http://bit.ly/1KWASmU
Golden Age of Islam: http://bit.ly/1RuWDOe
Goldsborough Nuclear Bomb Accident: http://bit.ly/1lGGGPS
Grasping Large Numbers: http://bit.ly/1o7Yasq
Growth rate exponential rebuttal: http://bit.ly/1Kz8ku1
Growth rates general: http://bit.ly/1xpKMGG
Habitability of M Dwarf systems: http://bit.ly/1UJoI0E
Habitable zones: http://bit.ly/1P08T6C
History – Middle Ages: http://bit.ly/Zofa5W
History of Earth on 24-hour clock template: http://bit.ly/1WcD9Rf
Human evolutionary timeline: http://s.si.edu/1uwCbl7
Human population growth rates: http://bit.ly/1PPaXiu
Human/Wolf relationship: http://bit.ly/1thczBT
Ice Age Land Bridge: http://bit.ly/1dhlqmT
Islam: http://bit.ly/1UmEwH3
Iron reserves: http://bit.ly/206qNtm
Kepler statistics: http://bit.ly/1H3HJVw
Leonardo Da Vinci: http://bit.ly/1dxK021
M-dwarf lifetimes: http://bit.ly/1GBvbtO
Microprocessor history: http://intel.ly/1oyLMzJ
Mona Lisa: http://bit.ly/1XUxYTA
Moon statistics: http://go.nasa.gov/2jhyP67
Moore's Law: http://bit.ly/1i7nJeq
Most Common Recent Ancestor: http://bit.ly/1KlDdDC
Natural disasters: http://bit.ly/1S4UEx2
Natural gas reserves: http://bit.ly/1iyyQaU
Neanderthals and Denisovans: http://bit.ly/10CTjbL
Neutron star spin rates: http://bit.ly/1WcNyY7
Newton prism: http://fla.st/1N1mEBI
Nuclear Incident in 1961: http://bit.ly/1lGGGPS
Oil source as single celled life: http://nyti.ms/1PPaXiy
Oil use: http://bit.ly/1AeSIrb
Oregon Trail: http://bit.ly/1p7oPLH
Oxygenation of oceans and atmosphere: http://bit.ly/2adNMAV
Photosynthetic cells: http://bit.ly/1Ch4Ijc
Photovoltaic coverage of Earth: http://bit.ly/1wwgr6v
Planck scale to decimal place: http://bit.ly/1WcDbJ8
Plant height limitations: http://bit.ly/1M9OLOo
Plant material for car gas: http://bit.ly/1Kz8wcK

Planetary dynamics: http://bit.ly/1OheVky
Radio communications travel distance: http://bit.ly/13onzcd
Search for exoplanet life: http://bit.ly/1ySMmNn
Size of Earth comparisons: http://bit.ly/1NuKTXl
Size of Universe comparisons: http://bit.ly/1XskZZ0
Solar cell efficiency: http://bit.ly/1C4mNxq
Solar Facts: http://bit.ly/1lb26Wa
Space resources: http://bit.ly/1MkNGoU
Size of the solar system: http://bit.ly/1WcNGXo
Star spectral classifications: http://bit.ly/1kw9Qbj
Star Trek: The Next Generation: http://bit.ly/1nczVOD
Steel use: http://1.usa.gov/1w6qrre
Stress Controlling Genes: http://go.nature.com/2guTuza
Super-Earth range of potential: http://bit.ly/IPMmvo
Super-Earth comparison to Earth: http://bit.ly/IPMmvo
Super-Earth surface gravity: http://bit.ly/1g1jqjh
Supernova rates: http://bit.ly/1OPPPcG
Sumerian civilization: http://bit.ly/1FeJUpf and http://bit.ly/1ySl0sH
The Great Oxygenation Event: http://bit.ly/2adNMAV
The late heavy bombardment: http://bit.ly/1MNig4Q
Time to Mars: http://bit.ly/1GBvEMx
Travel to other star systems: http://bit.ly/1Rw1meP
Types of civilizations: http://bit.ly/1MNhODO
Vestigial organs: http://bit.ly/1XT4PrU
Virtual reality devices: http://bit.ly/1g4Dpht
Wood charcoal: http://1.usa.gov/1w6qrre
Year Without a Summer: http://bit.ly/1hE9cGw

ABOUT THE AUTHOR

Mathew has been exploring the boundaries of what it means to be human since voluntarily stepping in wet cement while on the way to his first kindergarten class. The adventures and lessons learned since then have only become more unexpected and profound.

Mathew studies physics, astronomy, and planetary habitability. He regularly consults with studios and documentary projects on the topics found in this book. His goal is to raise everyone's awareness on why our existence as a civilization is so precious.